AQA
GCSE PHYSICS

Steve Witney, Barbara Drozdowska, Peter Maile

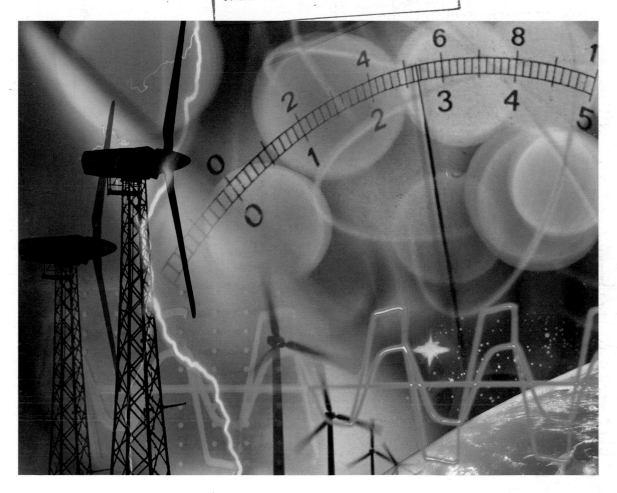

Hodder & Stoughton
A MEMBER OF THE HODDER HEADLINE GROUP

Photo acknowledgements

The publishers would like to thank the following individuals, institutions and companies for permission to reproduce photographs in this book. Every effort has been made to trace ownership of copyright. The publishers would be happy to make arrangements with any copyright holder whom it has not been possible to contact:

Action Plus (69, 82, 84 bottom, 88 bottom, 90 top, 94, 95 top, 170) Andrew Lambert (15, 26 bottom); Associated Press (180 top); Bruce Coleman Ltd (65, 71 all, 87); Corbis (81, 88 top, 147 top, 151, 177 top right, 185 both, 186 Niels Bohr); Hodder & Stoughton (92, 190); Holt Studios (51); Houghton's Horses (90 bottom); Life File (14, 20 bottom, 22, 26 top, 50, 55 top, 57, 89); Hulton Archive (186 top); Robin Marshall (186 Ernest Marsden); Natural History Museum (142 bottom left, 143 top); PA Photos (126); Phillip Harris (55 bottom); Ruth Nossek (120); Science Museum (1); Science Photo Library (2, 3 both, 4, 5, 20 top, 40, 42, 43, 45, 48, 58, 84 top, 95 bottom, 96, 111, 134, 140, 142 top right & bottom right, 143 left, 145, 146, 147 bottom, 149, 150 both, 152, 176 all, 177 top left, middle & bottom, 179, 180 bottom left, 181 bottom); Telegraph Library/Getty Images (43, 56); Wellcome Trust (47, 180 bottom right, 181 top).

Orders: please contact Bookpoint Ltd, 130 Milton Park, Abingdon, Oxon OX14 4SB.
Telephone: (44) 01235 827720. Fax: (44) 01235 400454. Lines are open from 9.00–6.00, Monday to Saturday, with a 24 hour message answering service.
You can also order through our website www.hodderheadline.co.uk

British Library Cataloguing in Publication Data
A catalogue record for this title is available from the British Library

ISBN 0 340 84779 4

First published 2002
Impression number   10   9   8   7   6   5
Year   2008   2007   2006   2005   2004

Copyright © 2002 Steve Witney, Barbara Drozdowska, Peter Maile

Cover illustration by Sarah Jones at Debut Art
Typeset by J&L Composition, Filey, North Yorkshire.
Printed in Italy for Hodder & Stoughton Educational, a division of Hodder Headline Ltd, 338 Euston Road, London NW1 3BH.

Contents

About this book

The contents

The contents of this book are designed to cover all aspects of the knowledge and understanding required by the AQA GCSE specifications in Physics (Co-ordinated) and Physics (Modular).

The subject content required by the KS4 Double Award specification for Physical processes attainment target is produced in a format identical to that used in the Hodder and Stoughton textbook *AQA GCSE Science*. This core material is supplemented by the additional subject content required for the specification in GCSE Physics.

What is in each chapter?

At the beginning of each chapter is a list of **Key Terms**. Where used for the first time, these key terms are emboldened. Some of the key terms are coloured. These are the extra terms you will need to know if you are going to be entered for the Higher tier papers in the final examination. All the key terms together with their meanings are also found in the **Glossary** on page 196.

The contents of each chapter are divided into several sections. Each section concentrates on one topic. A symbol at the start of each section shows clearly which topic from the co-ordinated and modular courses is being targeted.

You will see a number of **Did you know?** boxes throughout each chapter. You will not have to learn the information in these boxes, but they are there to give extra interest to the topic.

At the end of various sections, you will find a number of **Topic Questions**. Because the topic questions have been designd to produce answers that you could use as a set of revision notes, it is recommended that you write down the questions as well as the answers. The questions written on a yellow background are the more demanding questions, expected to be answered if you are a grade B/A/A★ student. Don't worry if you have to re-read the topic again when you try to answer these questions. This will help you to learn the work.

At the end of each chapter is a **Summary**. The summary provides a brief analysis of the important points covered in the section.

Completing each chapter are some **GCSE questions** taken from past AQA (SEG) or past AQA (NEAB) examination papers. The questions written on a yellow background are the more demanding questions expected to be answered if you are a grade B/A/A★ student. Answering the GCSE questions will help give you an idea of what is wanted when you take your final science examination. Again, do not worry if you have to go back to read the work again. The examination questions may well test you on knowledge not included in the particular chapter. Don't worry – look through the other chapters to find the extra information you need to complete our answer.

Specification Matching Grid

Chapter	Section	Content	Co-ordinated	Module	Section
			AQA specification references		
			Co-ordinated	Modular	
1 Electricity and magnetism	1.1	Electric charge	10.5	10	11.3
	1.2	Circuits	10.1	10	11.1
	1.3	Energy and power in a circuit	10.2	10	11.5
	1.4	Mains electricity	10.3	10	11.4
	1.5	Paying for electricity	10.4	09	10.2
	1.6	Electromagnetic forces ✓	10.25	10	11.2
	1.7	Electromagnetic induction	10.26	10	11.6
	1.8	Control in circuits	10.6	23	14.1–14.6
2 Forces and motion	2.1	Speed, velocity and acceleration	10.7	11	12.1
	2.2	Force and acceleration	10.8	11	12.2
	2.3	Frictional forces and non-uniform motion	10.9	11	12.2
	2.4	Turning forces	10.10	24	15.1,15.2
	2.5	Momentum	10.11	24	15.4
	2.6	Circular motion	10.12	24	15.3
3 Waves	3.1	Characteristics of waves	10.13	12	13.1
	3.2	The wave equation	10.13	12	13.1
	3.3	The electromagnetic spectrum	10.14	12	13.2
	3.4	Optical devices	10.15	23	14.7
	3.5	Sound and ultrasound	10.16	12	13.5
	3.6	Seismic waves	10.17	12	13.6
	3.7	Tectonics	10.18	24	15.5
4 The Earth and beyond	4.1	The solar system	10.19	11	12.4
	4.2	The wider Universe	10.20	11	12.5
5 Using energy and doing work	5.1	Thermal energy transfers ✓	10.21	09	10.1
	5.2	Efficiency ✓	10.22	09	10.3
	5.3	Energy resources	10.23	09	10.4
	5.4	Work, power and energy	10.24	09/11	10.2,12.2, 12.3
6 Radioactivity	6.1	Types, properties and uses of radioactivity	10.27	12	13.3
	6.2	Atomic structure and radioactivity	10.28	12	13.4
	6.3	Half life	10.27	12	13.3
	6.4	Nuclear Fission	10.28	12	13.4

Ideas and evidence in Science

You will find that many sections contain information which is marked with a bell and a vertical stripe in the margin. This is material to support the 'Ideas and Evidence in Science' part of your course. It will provide you with information about:

- how scientific ideas were developed and presented,
- how scientific arguments can arise from different ways of interpreting the evidence,
- ways in which scientific ideas may be affected by the contexts in which it takes place (for example, social, historical, moral and spiritual) and how these contexts may affect whether or not ideas are accepted,
- the problems science has in dealing with industrial, social and environmental questions, including the kinds of questions science can and cannot answer, uncertainties in scientific knowledge, and the ethical issues involved.

Each of the 'Ideas and evidence' contexts needed for whatever course you are following is included in this book. A guide to these contexts and whether they are needed for Core or Higher tier is given in Table 1.1 below.

Table 1.1   Contexts for the delivery of 'Ideas and evidence' in Physical Processes

Section	DA	Core/HT	Context
1.8	✗	core	Advantages and disadvantages of advanced electronic systems
3.3	✓	core	The dangers or possible dangers of exposure to different types of electromagnetic radiation and measures that can be taken to reduce such exposure
3.7	✓	core	Why the accurate prediction of earthquakes and volcanic activity is difficult
3.7	✓	core	Why Wegener's theory of Continental Drift took a long time to be accepted
4	✓	core	How scientists have tried to discover whether there is life elsewhere in the Universe
4	✓	HT	Why the theories of the origin of the Universe have to account for the 'red-shift'
5	✓	core	The advantages and disadvantages of using different energy sources to generate electricity
5	✓	HT	How different energy sources compare financially and economically in the generation of electricity
6	✓	core	How the Rutherford and Marsden scattering experiment led to the replacement of the 'plum-pudding' model of the atom with the present model of the atom

Chapter 1

Electricity and Magnetism

1.1 Electric charge

Co-ordinated	Modular
10.5	Mod 10 11.3

Electrostatics

If a balloon is rubbed on a cloth, the balloon will stick to the wall. The attraction between the balloon and the wall is caused by **electrostatic forces**.

Figure 1.1 ▶

Figure 1.2
Each hair has the same charge, so they repel each other

Friction between the balloon and the cloth moves **electrons** from some atoms in the cloth. These electrons transfer to the balloon from the cloth and so the balloon is negatively **charged**. The cloth is positively charged as it has lost electrons. Charged objects affect each other. The rules are:

- Objects with the *same* charge *repel*.
- Objects with *opposite* charges *attract*.

Electricity and Magnetism

The negative charge on the balloon repels electrons on the wall surface. Close to the balloon the wall becomes positively charged. As the balloon is light it is attracted to the wall. The balloon stays on the wall until its negative charge leaks away to the air or to the wall.

Figure 1.3 ▶
Positive charges stay on the wall surface

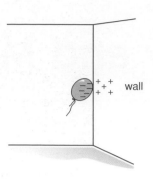

Charges at work

The sparks seen when clothing is removed quickly are due to electrostatic charges caused by friction. Clothes in a tumble dryer often become charged by friction causing 'static' clicks as the clothes are separated. Walking on a carpet can cause a person to get an electrostatic charge. This can result in the person getting an electric shock when they touch a metal door handle.

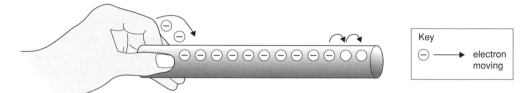

Figure 1.4 ▶
A conducting metal rod

This is because metals have atoms whose outer electrons are very free to move. Should some of these **free electrons** be rubbed off then other electrons easily take their place, producing a flow of electrons. Substances, such as metals, in which electrons can flow easily are good **conductors** of charge. Plastic, polythene and rubber conduct badly and are called **insulators**. If electrons are rubbed off an insulator, then other electrons do not flow to take their place. When electrons are rubbed onto an insulator they do not move away. A static charge builds up.

Figure 1.5 ▶
A large Van der Graaff generator

Insulators and insulated objects can become very highly charged. The very high charge is not able to flow.

Problems created by a large charge
Lightning

Figure 1.6 ▶
Lightning between oppositely charged clouds

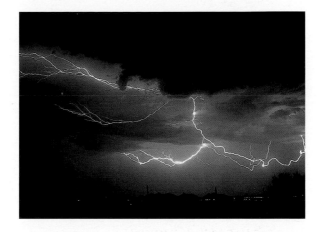

If a cloud with a large charge passes close to tall objects on the ground, then the charge can run to earth causing a lightning strike.

Figure 1.7▶
A lightning strike between a charged cloud and a tree

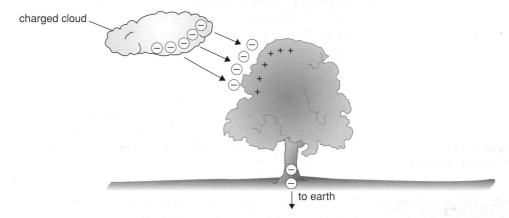

Many trees are damaged by lightning as they are often the tallest objects on the ground. The air particles between the charged cloud and the tree are **ionised**. Some air particles are stripped of electrons to become positive ions; others gain electrons to become negative ions. These ions allow the electrons to move quickly from the cloud to the tree during the discharge. The flash occurs as the air particles rejoin to their missing charge.

High voltage cables

High voltage cables are bare metal wires conducting large amounts of charge. They are held far above the ground by insulated pylons. Getting too close to such a cable is dangerous. An electric discharge can flow to the close object when the air stops insulating the cable. As the charge moves, a large **electric current** flows to earth.

Figure 1.8
The insulated supports are clearly visible on these pylons

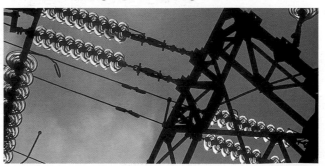

A kite touching the cables will complete an electric circuit. There is now a path for the current to earth through the person. The large flow of charge can kill the person holding the kite. Even a small current can stop the heart muscle working.

Figure 1.9

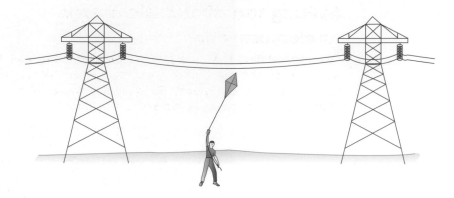

Did you know?

The lightning conductor was invented by Benjamin Franklin. He flew a kite during a thunderstorm and saw the line react as charge ran down it to earth. This gave him the idea for the lightning conductor. A Russian scientist copying Franklin's experiment died as the charge was conducted through him to the ground when he held a similar kite.

Delivering fuel

Figure 1.10
Tanker delivering fuel safely

As large tankers deliver petrol to a garage or aviation fuel to an aircraft, the movement of liquid in the pipe causes the pipe to be charged by friction. A wire connecting the pipe nozzle to the ground prevents a large charge forming on the pipe, because any charges leak away along the wire to the **earth**. If a large charge is formed on the pipe then a discharge spark to an uncharged object could ignite the petrol vapour.

Fine powders

In some industries, such as flour making, the friction between fast moving powders can cause the build up of electric charges. If the charges are not discharged safely, they can cause the powders to explode.

Making use of electric charge

An electrostatic paint spray gun

An electrostatic paint spray gun like the one in Figure 1.11 charges the paint droplets as they leave the gun. The droplets repel each other forming a fine spray. This coats the objects evenly with paint. Less paint is wasted if the object is given an opposite charge, as all the paint is attracted to the object.

Figure 1.11
An electrostatic paint spray gun

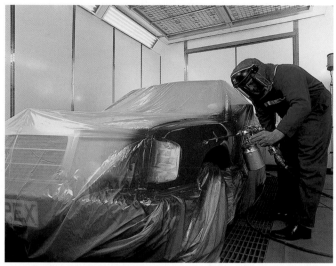

Electrostatic smoke precipitator

The burning of fossil fuels, such as coal, in power stations and factories pollutes the atmosphere not only with waste gases but also with smoke. Smoke consists of tiny particles of solid material. The smoke can be removed from the waste gases before they pass into the atmosphere by using the principles of electrostatics in a device called a smoke precipitator (Figure 1.12).

On their way up through the chimney the waste gases pass through a negatively charged metal grid. The smoke particles become charged as they pass by the grid and are repelled by the similar charge on the grid. The large collecting plates lining the chimney have an opposite (positive) charge to that on the grid. The smoke particles are attracted to the oppositely charged metal plates connected to earth. The smoke particles lose their charge and drop to the bottom when the plates are tapped. The waste gases are then free of smoke particles.

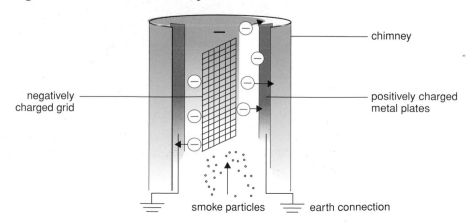

Figure 1.12
An electrostatic smoke precipitator

The photocopier

Substances such as selenium, arsenic and tellurium are called photoconductors because they are electrical insulators in the dark but electrical conductors in the light.

Figure 1.13 ▶
A photocopier

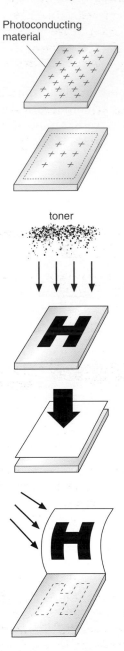

Photoconducting material

toner

1. In a photocopier there is a plate that consists of a layer of photoconducting material on a thin metal backing sheet. At the start of a copy cycle this plate is given a positive charge.

2. The page to be copied is lit with a strong light and an image of the page forms on the surface of the charged plate. The white areas of the page light up the photoconducting layer so it becomes conducting and the charge leaks away to the metal backing. The black areas of the page are left as an image in the positive electric charges on the belt.

3. Negatively charged toner powder is spread over the plate and is attracted to the parts of the plate still positively charged (the dark parts of the original copy).

4. A blank sheet of paper is given a positive charge and rolled over the belt where it attracts the negatively charged toner.

5. Rollers then apply heat and pressure to make the toner stick to the paper.

Electric current

An **electric current** is the movement of charged particles. In a metal wire, the electric current is the flow of electrons. Electrons always move from a place where there are too many electrons to a place lacking electrons. When a wire is connected to the terminals of an electrical source, the electrons travel from the negative end of the source, along the wire to the positive end or terminal of the source. This is the same as thinking of a flow of positive charges from the positive terminal. Remember, positive charges do not move in a solid.

Only the outer electrons of metal atoms are free to move through a solid.

The electric current in a gas or liquid is a movement of positively and negatively charged particles called **ions**.

Figure 1.14

Flowing electrons make an opposite electric current

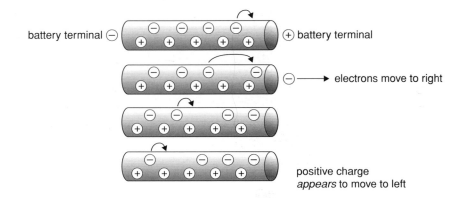

Did you know?

Scientists who carried out early work on electric current did not know about electrons. They thought a positive electric current flowed from the positive to the negative terminal. This idea is still used today. This idea is called conventional current.

Electrolysis

Certain chemical compounds conduct electricity when they are melted or dissolved in water. The electric current is formed by negatively charged ions moving to the positive **electrode** (**anode**), and positively charged ions moving to the negative electrode (**cathode**). The conducting liquid is called the **electrolyte** and the process is called **electrolysis**. When the ions reach the electrodes, simpler substances are given off at the electrode as gases, or deposited there as solids.

During the electrolysis of copper sulphate solution, the negative electrode becomes coated in a layer of copper, whilst the positive electrode loses copper (Figure 1.15).

In the electrolysis of slightly acidified water, hydrogen gas bubbles are given off at the negative electrode, and oxygen gas at the positive electrode.

The results of electrolysis reactions like these show that the mass or volume of a substance freed during electrolysis increases in proportion with:

- the current – doubling the current doubles the mass or volume
- the time for which current flows – doubling the time doubles the mass or volume.

7

Figure 1.15
How liquids conduct electric current

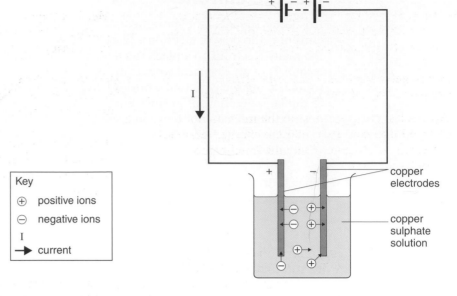

Key

⊕ positive ions

⊖ negative ions

I
→ current

copper electrodes

copper sulphate solution

Example: In the electrolysis of copper sulphate solution, 0.9 g of copper were deposited when a current of 1.5 A flowed for 30 minutes.

How much copper would there be for a current of 1.5 A for 90 minutes?

30 minutes	give	0.9 g
so 3 × 30 minutes	give	3 × 0.9 g
New mass	=	3 × 0.9 g
	=	2.7 g

Did you know?

Electroplating uses electrolysis. Metal spoons have an even coating of silver.

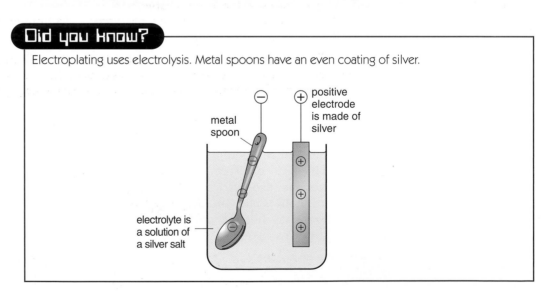

metal spoon

⊖ positive electrode is made of silver

electrolyte is a solution of a silver salt

Summary

◆ There are two types of electrical charge: negative and positive.

◆ An object becomes negatively charged when it gains **electrons**, and positively charged when it loses electrons.

◆ Two objects with a similar charge repel; two objects with opposite charges attract.

◆ **Electrostatic forces** can be used to explain the working of paint spray guns, photocopiers and smoke precipitators.

◆ A build up of charge can be dangerous.

◆ A large build up of charge on an object can cause a spark to jump between the object and any earthed conductor brought near to it.

◆ A charged object can discharge when it is connected by a conductor to earth (**earthing**).

◆ An electric current is due to the movement of charged particles. In a solid the charged particles are electrons. In a gas or liquid the charged particles are **ions**.

◆ Metals are good conductors of electricity because they contain free electrons that can move easily through the metal.

◆ **Electrolysis** is the process by which an electric current flows in a liquid to release simpler substances at the anode and cathode.

◆ The mass/volume of a substance deposited or released at the electrodes during electrolysis increases in proportion to the time and the current.

Topic questions

1 Copy and complete the following sentences using words from the box.

| attract | charged | electrons | friction | induces | positively |
| repel | rubbed | small |

A polythene rod is negatively _____ when it is _____ with a cloth. The cloth loses _____ to the rod. The cloth is _____ charged. The rod can attract _____ objects. The rod _____ opposite charges on the objects. Opposite charges _____ and like charges _____.

2 Match the term to each explanation.

discharge   earthing wire   electrolysis   electrons

a) Transfer a large charge safely to the ground.
b) Tiny particles with negative charge.
c) Rapid movement of electrons to a place where electrons are missing.
d) Electric current as ions move in a liquid.

3 A small polystyrene bead hangs on a nylon thread and is given a positive charge. Different rods charged by friction are brought near the bead. The movement of the bead is recorded in the table. Copy the table, tick each charge present on each rod.

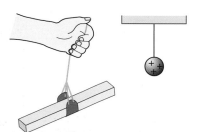

Material of rod	Bead movement	Charge on rod		
		Positive	Negative	Uncharged
Cellulose acetate	Repelled			
Ebonite	Attracted			
Perspex	Repelled			
Polythene	Attracted			
Steel	None			

9

4 Why are the nozzles of fuel pipes earthed?

5 A current flows through a solution of copper sulphate.

Which are the positive and negative ions?
Explain how you work out your answer.

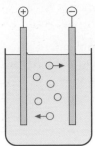

6 A lightning conductor is connected to this church tower, to prevent damage during a lightning discharge. The conductor is a metal spike joined by thick copper wire to a metal bar in the earth.

a) What happens to the air between the cloud and the tower before a lightning discharge occurs?

b) What can you say about the size of the current and the time it flows during a discharge?

c) Give reasons why the lightning conductor must:
   i) be made of metal
   ii) be made of a thick copper connecting wire
   iii) have its end buried in the ground.

7 In an electroplating process, 2 g of silver are deposited on a spoon when a current of 1 A flows for 30 minutes.

What would be the mass deposited if
a) 1 A of current flows for 15 minutes
b) 2 A of current flows for 30 minutes.

1.2   Circuits

Co-ordinated	Modular
10.1	Mod 10 11.1

A circuit is a closed pathway of wires and components which conduct the electrons from and back to the electrical source. Circuits are represented by circuit diagrams. They show all the components and wires using symbols.

Figure 1.16 shows the symbols with the components they represent.

Figure 1.16 ▼
Electrical components and their symbols

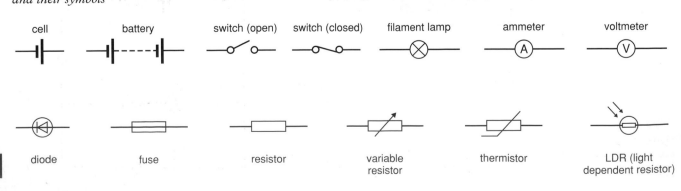

Series and parallel circuits

A series circuit is a single pathway for an electric current through a set of wires and components. A parallel circuit gives a number of pathways for the electric current.

In a series circuit with several lamps, removing one lamp breaks the circuit. All the other lamps go out. This also happens if a lamp 'blows'. Some sets of Christmas tree lights are wired in series. When one lamp burns out, all the lamps go out – with a long search to find the lamp that made all the others turn off.

In a parallel circuit with several lamps, removing one lamp does not stop the others working. Most lighting circuits are wired in parallel.

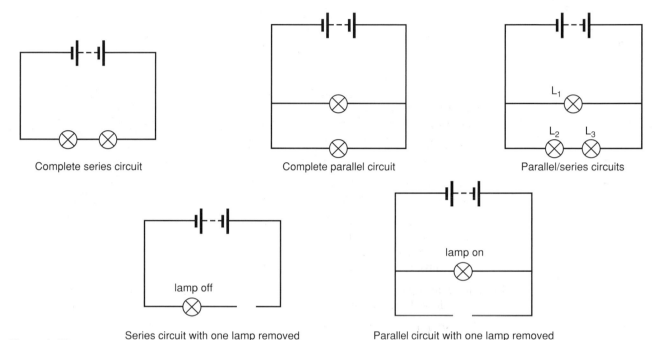

Complete series circuit   Complete parallel circuit   Parallel/series circuits

Series circuit with one lamp removed   Parallel circuit with one lamp removed

Figure 1.17

Electric current in series and parallel circuits

An electric current is the flow of electrons in a wire or the movement of ions in a fluid. Electric current, I is measured in units called **amperes** (A), often called amps. **Ammeters** are special instruments that measure the amount of electric current flowing in a circuit. Because ammeters measure the current in a circuit, they are always connected in series in the circuit.

The current in a series circuit is the same at all points in the path. The current in a parallel circuit divides between the different paths. It rejoins where the parallel paths meet. In a parallel circuit, the total current $= I_1 + I_2$.

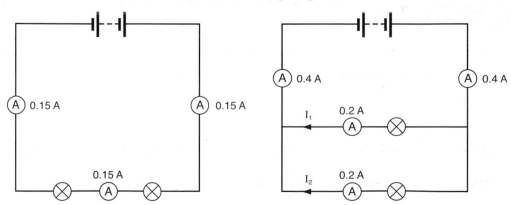

Figure 1.18
Current in series and parallel circuits

Potential difference (voltage)

The **potential difference** (p.d.) between two points in a circuit is the difference in electrical potential energy for a charge moved between those two points. The potential difference is often called **voltage** because a **voltmeter** measures the potential difference in units of **volts** (V). Because potential difference is the energy difference between two points, a voltmeter is always placed across the component in the circuit. So voltmeters are always connected in parallel with the component they are monitoring.

Potential difference – lamps in series

Lamps in series have the same current flow through each of them. This is because the same charge moves through each lamp. The charge gives up its energy in stages. Each stage sees a fall in the electrical energy of the charge. The greater the energy transferred to each lamp by each unit of charge, the larger the potential difference across the lamp.

Figure 1.19
Voltage drop and lamps in series

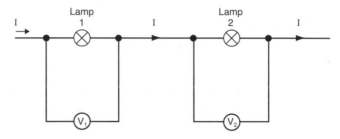

In a series circuit the total potential difference $= V_1 + V_2$.

Potential difference – lamps in parallel

Lamps in parallel have separate currents in each path. But the charges passing through each different path all have the same energy transfers. The potential difference across the set of parallel lamps is the same.

Figure 1.20
Current and lamps in parallel

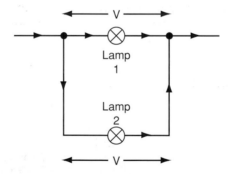

Potential difference – cells in series

The potential difference provided by cells connected in series is the sum of the potential differences of each **cell** separately – as long as they are connected together correctly, + to −.

Sometimes a single cell is wrongly called a **battery**. A battery is really a set of cells in series. A typical car battery has six cells, each 2V, with a total of 12V.

Figure 1.21
Potential difference of cells in series

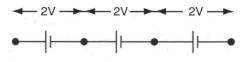

Total potential difference = 6 V

Resistance

Some devices let a large current flow through them. These devices have a low **resistance** to the flow of current. A device with a high resistance allows less current to flow through it when the same potential difference is applied. The amount of resistance is measured by seeing how much current flows for a certain potential difference. A typical circuit shows a component being tested for its resistance (Figure 1.22). The ammeter is always in series with the component, and the voltmeter in parallel with it.

The current is changed by the variable resistor to give a set of readings.

Current, in A	Potential difference, in V
0	0
0.5	2
1.0	4
1.5	6
2.0	8

By inspecting each pair of readings in the results table, it can be seen that $\dfrac{\text{potential difference}}{\text{current}}$ is a constant value (= 4).

A graph of potential difference against current shows a straight line through the origin. The slope of the graph is 4.

These results show that $\dfrac{\text{potential difference}}{\text{current}} = \text{constant}$

The constant is the resistance for that component and so

$$\begin{array}{ccccc} \text{potential difference} & = & \text{current} & \times & \text{resistance} \\ \text{(volt, V)} & & \text{(ampere, A)} & & \text{(ohm, } \Omega\text{)} \\ V & = & I & \times & R \end{array}$$

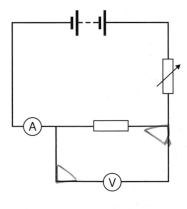

Figure 1.22

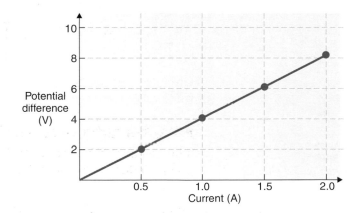

Figure 1.23

This resistance stays constant provided the component does not change its physical conditions (for example, the wire does not get hot). This relationship was first investigated by a scientist called George Ohm and the unit of resistance is named after him, the **ohm** (Ω).

Example: What is the potential difference across the 10 ohm wire in Figure 1.24?

$$\begin{aligned} \text{potential difference} &= \text{current} \times \text{resistance} \\ &= 3 \times 10 \\ &= 30 \text{ volt} \end{aligned}$$

Figure 1.24

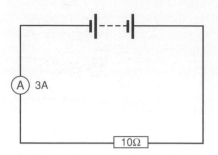

Resistances in series and parallel

Resistances in series all have the same current. As another resistance is connected in series the current decreases because the total resistance increases. When components are connected in series, the total resistance is the sum of their separate resistances.

Example:

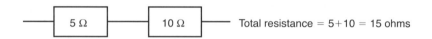

For resistances in parallel, each additional resistor connected in parallel allows another pathway for the current. So the total current increases, and the total resistance decreases. The total resistance is less than the smallest resistance in parallel.

Example:

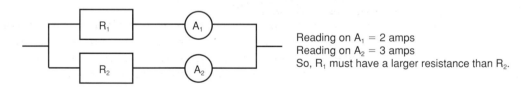

Reading on A_1 = 2 amps
Reading on A_2 = 3 amps
So, R_1 must have a larger resistance than R_2.

Circuits must not be overloaded, a temptation with adaptor plugs. The connecting wires carry too large a current, get very hot and can cause a fire.

Figure 1.25
Overloaded circuits cause fires

The connecting wire has a very small resistance. If it is put in parallel across a device, the wire takes most of its current in a 'short circuit' (Figure 1.26).

Figure 1.26
Short circuit prevents lamp from lighting

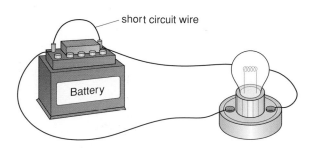

Wires of different materials have different resistances. Poor conductors have very high resistances so almost no current flows. But even materials that conduct do not all conduct to the same extent. Copper is a very good conductor – a copper wire will let a large current flow, it has very low resistance. But a steel wire of the same thickness and length, and with the same voltage across it, will not let such a large current flow. It is not such a good conductor – it has a higher resistance.

Figure 1.27
Graph showing differences in resistance for two wires

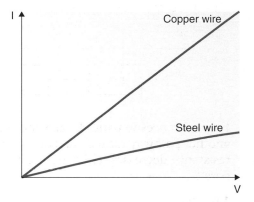

Conducting devices

1 Resistors

Resistors pass a certain amount of current for a given voltage. High value resistors let a small current pass, and low value resistors allow a large current to pass. Standard resistors have their resistance clearly marked and some value showing the maximum current allowed without overheating.

For a resistor at constant temperature the current – potential difference graph is a straight line through the origin (Figure 1.29). This means the resistance remains constant. The current is directly proportional to the potential difference.

Reversing the resistor, so the current flows through it in the opposite direction, does not change its resistance.

A variable resistor has a resistance that is changed from 0 ohms to its maximum value by sliding the connector across the wire coils. Sometimes the wire coils are in a small closed box and the connector turns.

Figure 1.28
A variable resistor

15

Figure 1.29
Current and potential difference for a resistor

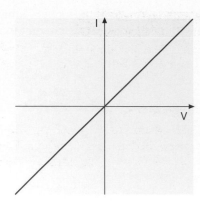

2 Diodes

Diodes are made of a non-metallic material called a semiconductor. A diode only conducts a current in one direction. If the direction of the current is reversed the diode does not let a current pass. The resistance of a diode depends on which way round it is connected. It has a very large resistance when connected in the reverse direction.

Figure 1.30
Diodes let current flow in one direction only

current flow –
forward connection

no current flow –
reverse connection

Figure 1.31
Lamp P lights but lamp Q will not as its diode is connected in the reverse direction to the current

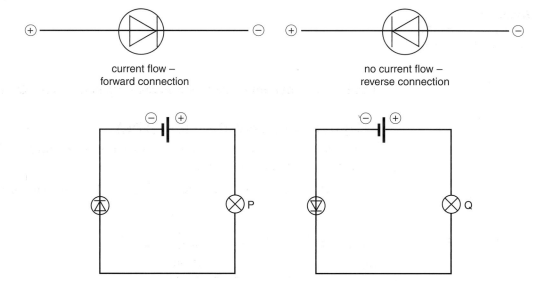

The graph linking current and voltage for a diode would look like Figure 1.32.

No current flows in the reverse connection as resistance is infinitely large, for normal potential differences.

Figure 1.32
Current and potential difference for a diode

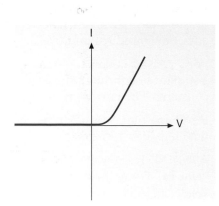

3 Filament lamp

A filament lamp uses a very hot wire to give light. As the voltage across the lamp increases, the light gets brighter. This is because the filament gets hotter. As the filament heats up its resistance increases. This means the current flow is less than expected.

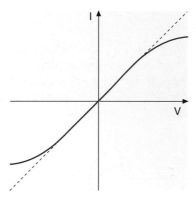

Figure 1.33
Current and potential difference for a filament lamp

The lamp has the same resistance increase when it is connected in the opposite way.

4 Light Dependent Resistors (LDR)

LDRs are also made of semiconductor material. They have a greatly reduced resistance when light falls on them.

The LDR is used in light operated circuits, such as security lighting.

Light level	Resistance, in ohms
dark	1 000 000
bright	500

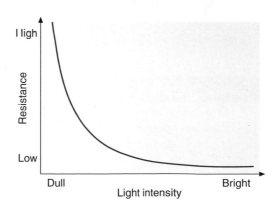

Figure 1.34
Variation of resistance with light intensity for an LDR

Figure 1.35
Symbol for an LDR

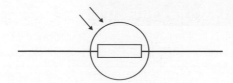

5 Thermistors

The **thermistor** is another semiconducting device. Its resistance decreases as the temperature increases. The drop in resistance causes a larger current to flow. The thermistor can be used in temperature sensing circuits.

Figure 1.36
Symbol for a thermistor

Summary

◆ For components connected in **series**:
  – the total **resistance** is the sum of their resistances
  – the same amount of **current** flows through each component
  – the total potential difference (voltage) is shared between the components.

◆ For components connected in parallel:
  – the total current in the circuit is the sum of the currents in each parallel component
  – there is the same potential difference (voltage) across each parallel branch.

◆ Current, potential difference (voltage) and resistance are related by the equation:

potential difference = current × resistance
  (volt, V)   (ampere, A)   (ohm, Ω)

◆ The current in a resistor (at constant temperature) is proportional to the potential difference across the resistor.

◆ For a filament lamp the resistance increases as the temperature of the filament increases.

◆ Diodes only allow a current to flow in one direction. For a diode the resistance is very high in the reverse direction.

◆ For a **light dependent resistor** (LDR) the resistance decreases as the amount of light shining on it increases.

◆ For a **thermistor** the resistance decreases as the temperature increases.

Topic questions

1 Match the circuit symbols with each component: ammeter, cell, fuse, lamp, variable resistor, switch.

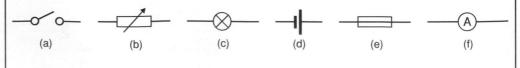

  (a)   (b)   (c)   (d)   (e)   (f)

2 Which switches have to be closed to make these lamps light?

   a) P and Q only
   b) R and T
   c) R only

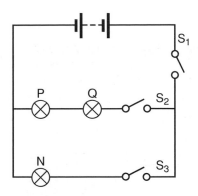
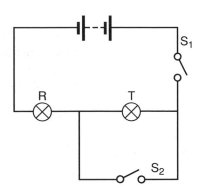

3 Which lamps are lit up when these circuits are switched on?

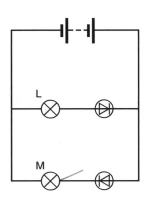
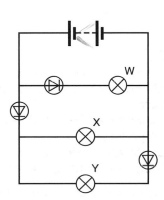

4 Copy and complete the following sentences.
The unit of resistance is the _____. A resistor has a _____ resistance provided the _____ stays the same. A filament lamp has an increase in its _____ as it gets _____ and brighter. The resistance of a _____ decreases as its temperature increases. A light dependent resistor or _____ has its highest resistance in the _____ .

5 a) Write down the equation linking current, potential difference and resistance.
   b) What is the potential difference across a 20 ohm resistor if a current of 0.5 A flows?
   c) What current flows through a 4 ohm resistor if the potential difference across it is 12 V?
   d) What is the resistance of a heater if the current flowing is 10 A and the mains voltage is 230 V?

6 a) What happens to the resistance of a filament lamp as it gets brighter?
   b) Why might a filament lamp 'blow' at the moment you switch it on?

7 a) When is the resistance of an LDR greatest?
   b) When is the resistance of a thermistor greatest?

1.3 Energy and power in a circuit

Electrical energy is used to work many machines; it is also used for heating lighting, televisions and communication systems. This energy is provided by a source of electricity, such as those shown in Figure 1.37.

(a)

(b)

Figure 1.37
Sources of electricity
a) Batteries
b) Power Station

A **cell** transfers its chemical energy to electrical energy when its terminals are joined by wires in a circuit. By joining a conductor from one end to the other, the negative charges can move to the positive terminal.

As the charge moves, an electric current flows through the circuit. The current stops flowing when the chemical reaction stops. This is when the cell is 'flat'. Some cells' chemical reactions can be renewed when they are recharged.

Figure 1.38 ▼
The current flows and the lamp is lit

Sometimes one cell is wrongly called a **battery**. A battery is really a set of cells in series. A car battery is a set of six cells in series. The battery recharges when the car engine turns.

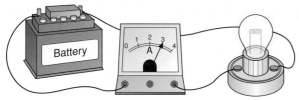

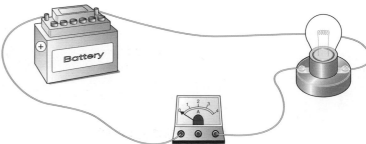

Figure 1.39
When the battery is 'flat', no current flows and the lamp stays off

As charge moves through the lamp, electrical energy is transferred to heat and light energy by the lamp.

The total amount of energy transferred depends on the total amount of charge moved and the potential difference of the source of electricity. The amount of energy is measured in a unit called a **joule** (J).

Figure 1.40
A battery with charger

Heating effect

Heating elements are usually made of special wire. They can transfer the electrical energy to heat energy efficiently without damaging the wire. Nichrome wire is frequently used in heating elements.

Figure 1.41
Some appliances designed to deliver heat energy

cooker

iron

Measuring electric charge

The amount of charge transferred in a circuit depends on the size of the current and the time for which it flows. The **coulomb**, C, is the unit of **electric charge**.

charge	=	current	×	time
(coulomb, C)		(ampere, A)		(seconds, s)
Q	=	I	×	t

Energy transferred by a charge

The more charge that is moved, and the bigger the potential difference across the terminals of the electrical source, then the larger the electrical energy delivered to the circuit components.

energy transferred	=	potential difference	×	charge
(joule, J)		(volt, V)		(coulomb, C)
E	=	V	×	Q

Potential difference, or voltage, gives the amount of energy transferred by one unit of charge. A voltmeter records 1 volt when 1 joule of energy is transferred by 1 coulomb of charge.

Example: A liquidiser is used from a mains supply at 230 volts. How much electrical energy is transferred to the liquidiser by a current of 1 A in 3 minutes?

charge	=	current	×	time		
	=	1	×	3	×	60
	=	180 C				

energy transferred	=	potential difference	×	charge
	=	230	×	180
	=	41 400 J		

Electric power

How quickly energy is transferred is as important as how much energy there is in the transfer. Walking slowly up a hill is difficult, but running up it is much harder. The same amount of energy is used each time but running uses the energy more quickly. It requires more **power**. Electrical power measures how quickly electrical energy is transferred by various devices to heat or light or movement, for example.

Figure 1.42
Electrical devices

Electricity and Magnetism

Power, whether muscle power or electrical power, is measured in a unit called the **watt** (W). A power of 1 watt is the transfer of 1 joule of energy in 1 second. Electrical power is related to the current and potential difference as:

$$\begin{array}{ccccc} \text{power} & = & \text{potential difference} & \times & \text{current} \\ \text{(watt, W)} & & \text{(volt, V)} & & \text{(ampere, A)} \end{array}$$

$$P = V \times I$$

The kettle shown in Figure 1.42 has a high power rating, 2000 W. It transfers electrical energy to heat energy quickly and so heats the water quickly. A **kilowatt** (kW) is 1000 W, so 2000 W can be written as 2 kW.

Example: A TV and video draw a current of 2 A from a 230 V power supply. What is the power rating of the TV and video?

$$\begin{array}{ccccc} \text{Power} & = & \text{potential difference} & \times & \text{current} \\ & = & 230 & \times & 2 \\ & = & 460\ \text{W} \end{array}$$

Summary

◆ In many electrical appliances electrical energy is transferred as heat.

◆ **Power** is the rate at which energy is transferred.

◆
$$\begin{array}{ccccc} \text{power} & = & \text{potential difference} & \times & \text{current} \\ \text{(watt, W)} & & \text{(volt, V)} & & \text{(ampere, A)} \end{array}$$

◆ For a given charge, the higher the potential difference the more energy is transferred, so
$$\begin{array}{ccccc} \text{transferred energy} & = & \text{potential difference} & \times & \text{charge} \\ \text{(joule, J)} & & \text{(volt, V)} & & \text{(coulomb, C)} \end{array}$$

◆ Charge, current and time are related by the equation:
$$\begin{array}{ccccc} \text{charge} & = & \text{current} & \times & \text{time} \\ \text{(coulomb, C)} & & \text{(ampere, A)} & & \text{(seconds, s)} \end{array}$$

Topic questions

1 Copy and complete the table for the quantity and its unit

Quantity	Unit
Current	
	Watt
	Joule
Resistance	
Potential difference	

2 What is the potential difference across these?
   a) torch lamp, R = 5 ohm, I = 0.25 A
   b) resistor, R = 75 ohm, I = 3 A
   c) heater, R = 45 ohm, I = 5 A

3 What is the power rating in watts for:
   a) a motor on a 125 V supply, taking a 2 A current?
   b) a kettle on a 230 V supply, taking a 10 A current?
   c) an electric oven on a 230 V supply, taking a 15 A current?

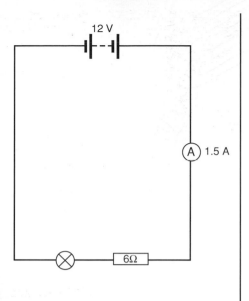

4 a) In the diagram what is the potential
   difference across:
   i) the resistor
   ii) the lamp
   b) What is the resistance of the lamp?

5 A 24 W set of Christmas lights has 20
   lamps in series worked from a 120 V
   supply. What is the current through the
   lamps? What is the potential difference
   across one lamp? What is the power
   transferred to each lamp?

6 A 12 V battery drives a motor taking a
   current of 2 A for 20 seconds. What is the
   electric energy transferred to the motor?

7 A 60 W lamp used on the mains supply
   of 230 V is switched on for 2 hours. How
   much electric charge was transferred in
   this time?

Co-ordinated	Modular
10.2	Mod 10 11.4

1.4   Mains electricity

Alternating current and direct current

A **direct current** (d.c.) always flows in the same direction, from a fixed positive
terminal to the fixed negative terminal of a supply. A cell or battery gives a constant
direct current (Figure 1.43). A cathode ray oscilloscope (CRO) shows how the
potential difference (voltage) changes with time.

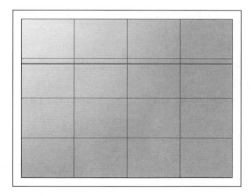

Figure 1.43
*The wave trace on a CRO for a d.c. supply
from a cell. The zero line is shown in red*

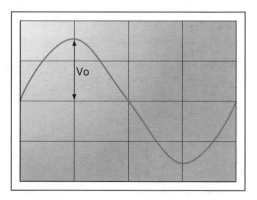

Figure 1.44
*The wave trace on a CRO for an a.c. supply.
The zero line is shown in red. The height of
the wave gives the peak voltage, V_O*

An **alternating current** (a.c.) in a circuit changes direction as the terminals of the
supply change from positive to negative.

When the CRO trace shows a potential difference below the zero line, the current is
in the opposite direction to that above the zero line (Figure 1.44). You can use a CRO
to measure the frequency with which the current changes direction. Frequency is
measured in **hertz** (Hz), giving the number of cycles in one second.

23

Taking measurements from a CRO trace for an a.c. supply

For the CRO trace in Figure 1.45, each vertical division represents 2 volts of potential difference. Each horizontal division represents 5 milliseconds of time. Therefore, this trace shows a maximum or peak potential difference of 6 volts and the time for one cycle of 40 milliseconds, giving the frequency as 25 hertz.

Figure 1.45

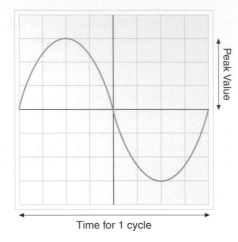

Time for 1 cycle

Peak value	=	3 divisions	×	2 volts per division
	=	6 V		
Time for one cycle	=	8 divisions	×	5 millliseconds per division
	=	40 milliseconds		
	=	0.04 seconds		
So, frequency	=	$\dfrac{1}{0.04}$		
	=	25 hertz		

Comparing CRO traces for two different a.c. supplies

Traces from two a.c. supplies were seen on a CRO, with the same settings for the vertical, potential difference reading and the horizontal, time reading.

Figure 1.46
a) 1 cycle takes 8 divisions
b) 2 cycles take 8 divisions

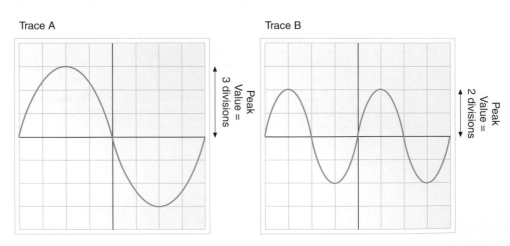

Trace A

Trace B

If the supply shown in trace A has a peak value of 6 V, then supply B will have a peak value of 4 V. The frequency of supply B is 2 times that of supply A. So if A has a frequency of 50 Hz, then B has a frequency of 100 Hz.

The mains electricity supply

In the United Kingdom, the mains electricity supply is about 230 volts. The supply provides an alternating current with a frequency of 50 hertz. This means the current flows one way, then the other, 50 times each second.

Safety devices and electrical appliances

Potential differences greater than 50 V and currents as small as 50 milliamps can be fatal to humans. Safety measures include the use of insulation and a ways of stopping or diverting the current.

The three pin plug

These days new appliances have plugs already attached. On older appliances, plugs need to be fitted. Care must be taken to connect the wires from the appliances to the correct pins in the plug. The colour coded wires help in this. The green/yellow striped earth wire protects the user should the live wire loosen and touch the appliance. By **earthing** the appliance with a wire leading directly to the ground, the current from the loose live wire flows to earth along the earth wire and not through the person touching the appliance!

Figure 1.47
Correct wiring in a three-pin plug

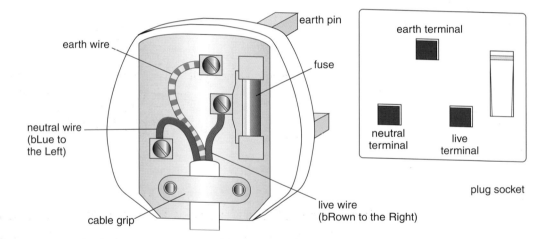

- The outside case of the plug is made of plastic or rubber, because these are good insulators.

- Pushing the plug into a socket connects the pins of the plug to the terminals of the socket.

- The live terminal of the mains supply alternates between a positive and negative voltage when compared with the neutral terminal. The neutral terminal stays at a voltage close to zero.

Electricity and Magnetism

Fuses

A **fuse** is connected to the live wire before it joins the appliance. It is there to shield the appliance from too large a current. If a fault causes a larger current to flow through the appliance than that marked on the fuse, then the fuse melts and so breaks the circuit.

Figure 1.48
Fuses

Each appliance has a power rating such as

> Power 460 W
> Supply 220-240 V
> ~ 50Hz
>
> (~ indicates an a.c. supply)

The working current for the appliance is found using

$$P \quad = \quad V \times I$$
$$460 \quad = \quad 230 \times I$$
$$I \quad = \quad 460/230$$
$$= \quad 2 \text{ A}$$

Figure 1.49
The power rating label on the side of an electric jig-saw

This is the normal current for the device and any larger current can destroy it. A 3 A fuse would allow a normal working current to flow and protect the appliance from larger currents. A 13 A fuse would allow a dangerously high current to flow and still not 'blow'. So it is important to use the correct size of fuse. A fuse does not protect the person using the appliance. It can take 1 to 2 seconds for a fuse to melt – enough time for the user to receive a fatal electric shock.

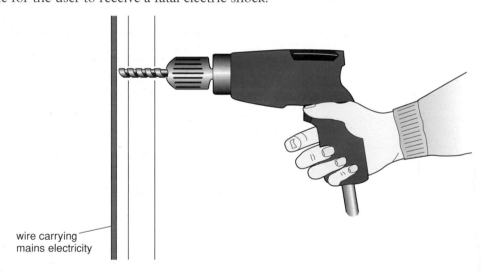

Figure 1.50 ▶
The current will flow through the person to the ground if there is no earth wire connected to the casing of the drill

wire carrying
mains electricity

Double insulation

Insulation is the first protective barrier for the user. Wires have an outer plastic or rubber coat. Many appliances have a plastic casing and do not have the earth wire. A loose live wire does not conduct through the plastic casing. The insulation of both the casing and the wires is shown by a double insulation label ▣ .

Circuit breakers

A residual current device, RCD, is a fast acting circuit breaker. It stops current flowing in less than 0.05 seconds. If the RCD detects any difference between the current in the live and neutral wires it will break the circuit.

Modern domestic wiring methods use a set of miniature circuit breakers (mcb) instead of a fuse box. Each mcb protects one circuit from too large a current overheating the wires and starting a fire. Once the fault is corrected, the mcb is reset. (see section 1.6). Many older houses have individual fuses for each wiring circuit, held in a fuse box. These are slower in their action and take longer to repair.

Summary

◆ **Alternating current** (a.c.) changes direction but **direct current** (d.c) flows in one direction only.

◆ In the UK the mains supply is at 230 V with a frequency of 50 hertz.

◆ Traces on a CRO can be used to compare the potential difference for a.c. and d.c. supplies.

◆ All electrical circuits should include safety devices because even a very low current could be fatal.

◆ The three-pin plug correctly wired contains a fuse and an earth wire.

◆ The earth wire is connected to the metal casing of many appliances. Other appliances use double insulation.

◆ Electrical circuits can be protected by circuit breakers.

Topic questions

1 What are the differences between:
   a) alternating current
   b) direct current?

2 Copy and complete the gaps in the table to provide information about the wiring of a three-pin plug.

wire	Colour
	Yellow/green
neutral	
	brown

3 a) To which part of an electric fire would an earth wire be fitted?
   b) How does an earth wire stop someone getting an electric shock?

4 Explain two ways a circuit breaker is more efficient as a safety device than a fuse?

5 Fuses for household appliances can be bought in these values 2 A, 3 A, 5 A, 10 A and 13 A.

   Domestic electricity is supplied at 230 V. Which fuse is correct for each of the following appliances?

Appliance	Power rating
Lamp	150
Television	200
Iron	800
Kettle	2400

1.5   Paying for electricity

Co-ordinated	Modular
	Mod 09
10.4	**10.2**

There are many examples of electrical appliances that transfer electrical energy into other forms of energy, such as:

● heat (thermal energy)
● light
● sound
● movement (kinetic energy).

The cost of using an electrical appliance depends upon the particular energy transfer.

Electricity and Magnetism

Heating appliances are expensive to run. The cost also depends on how quickly the energy transfer takes place. This is the electrical power delivered.

Figure 1.51
Some appliances designed to transfer energy from one form to another

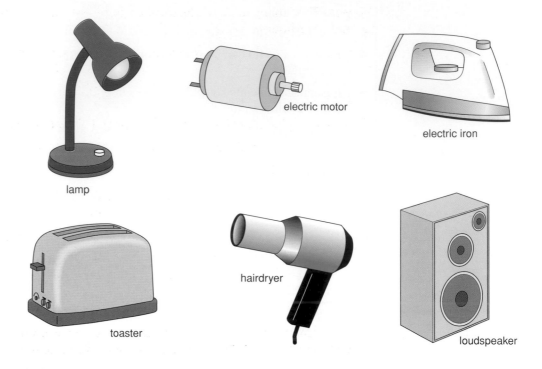

Appliance	Useful energy transfer
Toaster	Electricity to heat
Hairdryer	Electricity to heat and movement
Electric motor	Electricity to movement
Lamp	Electricity to light
Loudspeaker	Electricity to movement and sound
Electric iron	Electricity to heat

Figure 1.52

Did you know?

1 kWh is equivalent to 3.6 million joules of energy.

The power rating of an appliance is given in watts or **kilowatts**. This gives the amount of electrical energy transferred in 1 second by a working appliance (1 kW = 1000 W).

Electrical companies bill customers for electrical energy in special units of **kilowatt-hours** (kWh). One kilowatt-hour is the energy transferred when 1000 W is delivered for 1 hour.

Figure 1.53
Electricity meter showing two readings after an interval of time

February   May

These meters show that 57139−55652=1487 Units or kilowatt-hours of electrical energy was transferred.

1 Unit   =   1 kilowatt hour   =   1000 watt for 1 hour

$$\text{electrical energy transferred} = \text{power} \times \text{time}$$
$$\text{(kWh)} \qquad\qquad \text{(kW)} \qquad\qquad \text{(h)}$$

Example: How many units of electrical energy were supplied to a 100 W lamp left on at night for 8 hours?

$$
\begin{aligned}
\text{energy transferred} &= \text{power} \times \text{time}\\
&= 0.1 \times 8\\
&= 0.8 \text{ kWh}\\
&= 0.8 \text{ Units}
\end{aligned}
$$

The cost of 1 Unit (1 kWh) varies. If the price is 7p per Unit:

$$\text{total cost} = \text{number of Units} \times \text{cost per Unit}$$

So for lamp left on all night:

$$
\begin{aligned}
\text{total cost} &= 0.8 \times 7\\
&= 5.6\text{p}
\end{aligned}
$$

Heating appliances are the greatest energy consumers. An immersion heater, rated at 3000 W and used for 8 hours at 7p per Unit:

$$
\begin{aligned}
\text{total cost} &= \text{energy transferred} \times \text{cost per Unit}\\
&= 3 \times 8 \qquad\qquad\ \times\ 7\\
&= 168\text{p}
\end{aligned}
$$

The total amount of electrical energy transferred, measured in joules, is calculated using the same equation (see section 5.4) but with these units:

$$\text{energy transferred} = \text{power} \times \text{time}$$
$$\text{(joule, J)} \qquad \text{(watt, W)} \quad \text{(seconds, s)}$$

Example: A kettle with a power rating of 2500 W is left on for 5 minutes. How much electrical energy has been transferred in that time?

$$
\begin{aligned}
\text{energy transferred} &= \text{power} \times \text{time}\\
&= 2500 \times (5 \times 60)\\
&= 750\,000 \text{ J}\\
&= 750 \text{ kJ}
\end{aligned}
$$

This is a large number, so for commercial purposes the kilowatt-hour (kWh) is used.

Summary

◆ The power of an appliance is measured in watts (W) or **kilowatts** (kW). 1000 W = 1 kW

◆ The amount of energy transferred from the mains is measured in kilowatt hours or Units. 1 Unit = 1000 W delivered for 1 hour = **1 kilowatt-hour** (kWh)

◆ The amount of electrical energy transferred can be calculated using:

$$\text{energy transferred} = \text{power} \times \text{time}$$
$$\text{(joule, J)} \qquad \text{(watt, W)} \qquad \text{(second, s)}$$

or:

$$\text{energy transferred} = \text{power} \times \text{time}$$
$$\text{(kilowatt hour, kWh)} \quad \text{(watt, W)} \quad \text{(hour, h)}$$

◆ The readings on an electricity meter can be used to calculate the number of Units used.

◆ The cost of the energy used can be calculated using:

$$\text{total cost} = \text{number of Units} \times \text{cost per Unit}$$

1 How many units of electrical energy in kWh are used for a
   a) 100 W lamp for 10 hours?
   b) 200 W TV for 4 hours?
   c) 2000 W kettle for 5 minutes?

2 If the price for 1 unit is 7p. What is the running cost of each item in question 2?

1.6 Electromagnetic forces

Co-ordinated	Modular
10.25	Mod 10
	11.2

Magnetic repulsion and attraction

As long ago as 1000 BC it was known that certain rocks attracted iron objects. The mineral that did this was named magnetic after the region where it was first discovered. Pieces of this mineral would point to the north and could be used for navigation. These days **magnets** are manufactured. A freely-suspended magnet will line itself up along the north–south line of the Earth.

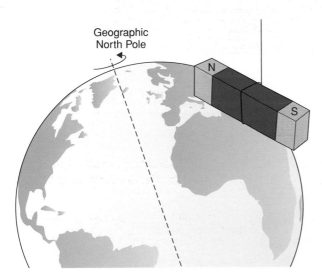

Figure 1.54
A bar magnet suspended in the Earth's magnetic field

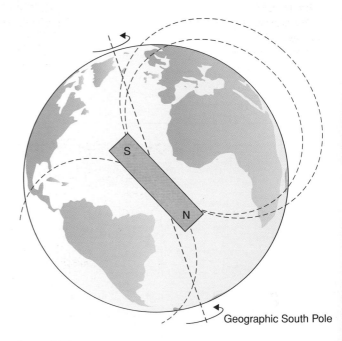

Figure 1.55
The theoretical bar magnet in the Earth that creates the Earth's magnetic field

Navigational compasses like this have been used since the eleventh century. Compasses work because the Earth behaves like a giant magnet with its north pole a little eastwards of the geographic south pole. The direction and strength of this magnet gradually changes over very long periods of time.

Experiments with magnetic forces show that:

- like magnetic poles repel
- unlike magnetic poles attract.

This is like the result for electrostatic charges. In magnetic substances, the atoms line up in groups to form tiny magnets. A larger magnetic force is produced when all the tiny magnets point the same way.

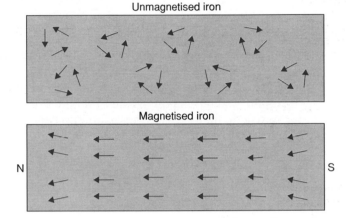

Figure 1.56
Arrangement of 'atomic magnets' in unmagnetised and magnetised iron

Magnetic fields

The metals iron, steel, nickel and cobalt are strongly attracted to magnets. Some magnetic substances can be made into permanent magnets. Temporary magnets can be made by passing a current along a wire.

The magnetic force is increased if the wire is coiled. Putting an iron bar in the coil strengthens the magnet even more. The coil of wire is called a **solenoid**. This is a simple **electromagnet**.

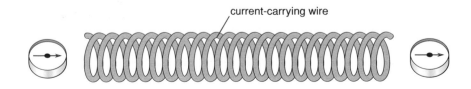

Figure 1.57
A simple electromagnet

Electricity and Magnetism

Magnets attract small iron filings and change the direction in which a compass needle points. The space around the magnets where the force acts is called the **magnetic field**. The shape of a magnetic field is shown by the iron filings. The magnetic force is strongest at the poles. A compass needle shows the direction of the magnetic lines of force.

Figure 1.58
The magnetic field around a) a straight wire, b) a solenoid

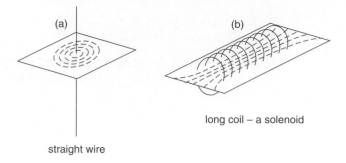

The direction of the magnetic field is shown by the pointing of the north pole of a plotting compass.

Increasing the number of turns (coils) of wire around an iron **core** increases the strength of an electromagnet. If the current in the wire is increased, an even stronger force is produced. Changing the direction of current reverses the magnetic poles. When the current stops, the magnetic field disappears. Any iron or steel objects held by the magnet fall off, as the iron core does not stay a magnet. (Steel and some other magnetic materials can be made into permanent magnets by putting them in an electric coil.)

Figure 1.59
The direction of the magnetic field around a) a straight wire, b) a solenoid

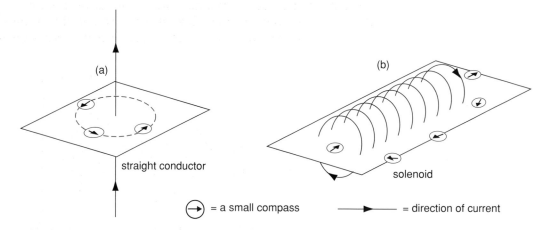

Using electromagnets

Large electromagnets are used to pick up steel and iron objects in a scrap yard or a factory. Small electromagnets are used in buzzers, bells, relay switches and miniature circuit breakers.

How a simple circuit breaker works

Figure 1.60 shows a simple **circuit breaker**. When a normal current flows in the circuit and around the coil of the electromagnet, the electromagnet is not strong enough to attract the iron bolt. The iron bolt is held in place by a spring. If the current increases due to a fault in the circuit, the electromagnet becomes strong enough to pull the iron bolt to it. This releases the plunger, which is pulled away from the contact switch by its own spring. The contact switch now opens, breaking the circuit, and the current stops flowing through the electromagnet. The electromagnet can no longer pull the iron bolt to it.

Once the fault is repaired, the reset button pushes the plunger that closes the switch. The iron bolt springs back into the slot in the plunger and this keeps the contact switch closed. A circuit breaker like this can be made to work at certain current values, protecting a circuit or device from having a current larger than the stated value.

Figure 1.60
A circuit breaker

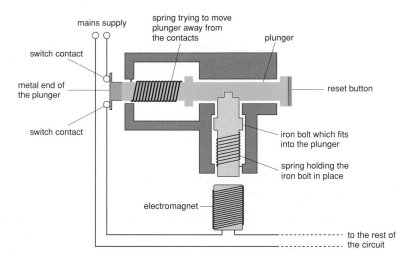

The motor principle

When a wire with a current flowing through it is placed between the poles of a magnet, on one side the magnetic field of the wire strengthens the magnetic field of the magnet, on the other side it weakens it. So one side of the wire has a stronger magnetic field than the other. The wire is pushed towards the weaker side of the field.

Figure 1.61
The motor principle

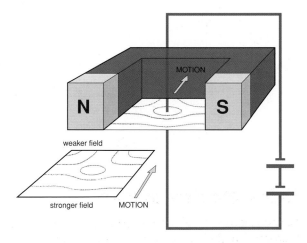

The force causing motion is greatest when the wire is at right angles to the field. No motion occurs if the wire is parallel to the field. Reversing the direction of the current, or the magnetic field, reverses the direction of motion. A stronger force is produced if the magnetic field strength or the current is increased.

If a coil of wire with a current flowing through it is placed in a magnetic field, the coil tends to turn. As one side of the coil moves up, the other goes down. If the coil is vertical, when the current direction changes, the coil will continue to turn. This is how a direct current electric motor works. A split ring ensures that the current flow changes direction at the right time.

Figure 1.62
A direct current motor

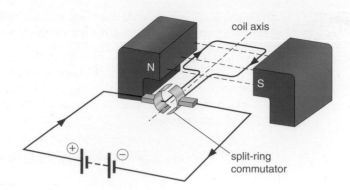

Summary

◆ A magnetic effect is produced by a current in a wire.

◆ An electromagnet is used in circuit breakers.

◆ A wire carrying current in a magnetic field experiences a force.

◆ This effect is made use of in d.c. motors.

Topic questions

1 Draw the magnetic field pattern for a:
   a) straight wire conductor
   b) solenoid.

2 Copy and complete the following sentences.

   In magnets, like poles _____ and _____ poles attract. A temporary magnet is made by _____ wire around an _____ bar.

3 The diagram shows an electromagnet in a relay switch when no current flows.
   a) Which of the contacts are joined together?

   The electromagnet is then turned on

   b) In which direction does the metal strip move?
   c) Which contacts are now joined together?

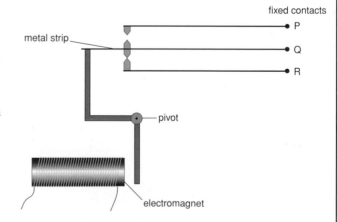

4 A student makes a simple motor with 10 coils of wire around a square plastic frame. The frame is on a spindle so it can turn between opposite poles of two magnets. The wire ends are either side of the spindle, touching metal contact strips connected to one cell.
   a) What happens if the magnetic poles are the same?
   b) What are the three different ways of making the coil turn faster?

Co-ordinated	Modular
	Mod 10
10.26	11.6

1.7 Electromagnetic induction

Just as moving charges cause a magnetic field, so a changing magnetic field about a conductor produces a current. This effect is easily shown by moving a wire through the magnetic field between the poles of a strong magnet. A potential difference is induced across the ends of the wire and a current is made to flow. This is **electromagnetic induction**.

Electromagnetic induction transfers kinetic (movement) energy to electrical energy. A current flows only when the wire moves in the magnetic field.

Figure 1.63
Electromagnetic induction

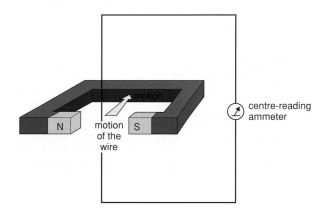

The dynamo principle

A coil of wire has a current induced when a magnet is moved into or out of the coil. Neither potential difference nor current is induced when the magnet is still. The current flows the opposite way when the magnet is moved in the opposite direction.

If the pole entering the coil is changed, the direction of current is reversed. The induced potential difference or current is increased by moving the magnet faster or using a stronger magnet or by having more turns on the coil cycle.

Figure 1.64
Electromagnetic induction in a coil

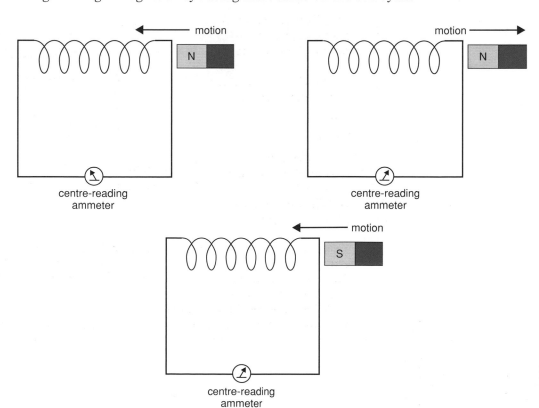

Figure 1.65
The effect of altering the pole on the direction of current

Electricity and Magnetism

This idea is used in making a **dynamo** to produce a current. It is not necessary to move a magnet in a coil to get a current induced. Rotating a coil of wire between the poles of a magnet will also induce a current to flow in the wire.

The induced potential difference or current is increased when:

- the coil rotates faster (or if the coil is stationary the magnet rotates faster).
- the area of the coil is increased.
- there are more turns on the coil.
- the strength of the magnetic field is increased.

Figure 1.66
A bicycle dynamo

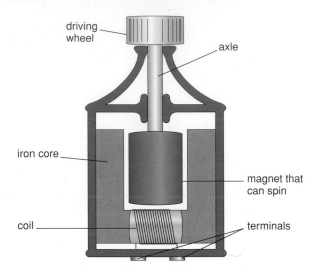

Figure 1.67
The dynamo principle

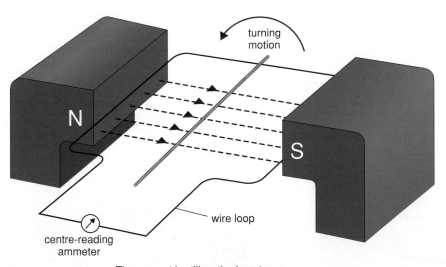

The magnet is still as the loop turns.
Each side cuts across the magnetic field.
An a.c. current is induced in the loop.

Simple a.c. generator

Figure 1.68
A simple a.c. generator

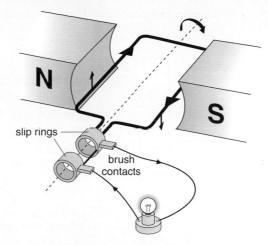

The a.c. generator has a fixed magnet and a coil that rotates. As the coil rotates a voltage is induced across the wire of the coil. Each end of the coil is connected to a conducting ring (slip ring) that also turns with the coil. The slip rings come into contact with two fixed carbon brushes. As the coil turns, the induced voltage changes direction for each half turn of the coil. The alternating current so produced passes via the slip rings and the brushes to the rest of the circuit.

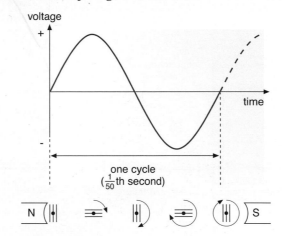

Figure 1.69
As the coil turns an a.c. wave is produced

Figure 1.70
A simplified mains a.c. generator

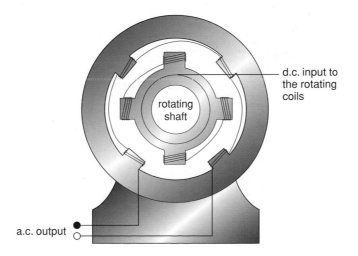

Other types of electricity generator use large stationary coils. A changing magnetic field is provided by rotating an electromagnet inside the stationary coils. The rotating electromagnet is usually driven by steam turbines using coal, oil, gas or nuclear energy as the energy source. The current produced in the stationary coils is an alternating current (a.c.).

The transformer principle

A changing magnetic field in a fixed coil will induce a current in a second fixed coil, if there is a magnetic link between the two coils. An iron core will provide such a link.

Figure 1.71
Obtaining an induced current

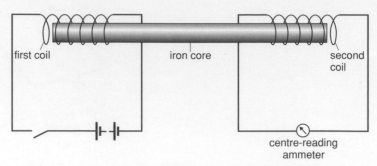

There is a current in the second coil only as the switch is closing or opening in the first coil. The magnetic field in both coils then changes.

Transformers and electromagnetic induction

Transformers use magnetic linking between two coils to step-up or step-down alternating voltages. They can only work using an alternating current. The changing magnetic field created by the alternating current in the **primary coil** causes an alternating current in the **secondary coil**. One of the reasons why mains electricity generators give an a.c. supply is to allow the use of transformers.

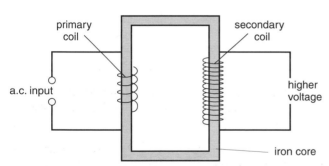

Figure 1.72
A step-up transformer

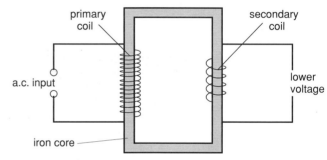

Figure 1.73
A step-down transformer

If the number of turns on the secondary coil of a transformer is more than on the primary coil, the output voltage is stepped up (increased). If the number of turns on the secondary coil is less than on the primary coil, the output voltage is stepped down.

This equation can be used to calculate the output voltage:

$$\frac{\text{voltage across primary coil(volt, V)}}{\text{voltage across secondary coil(volt, V)}} = \frac{\text{number of turns on primary}}{\text{number of turns on secondary}}$$

$$\frac{V_p}{V_s} = \frac{N_p}{N_s}$$

Example: A transformer is designed to step-down 230 V to 11.5 V. There are 1000 turns of wire on the primary coil. Calculate the number of turns of wire on the secondary coil.

Figure 1.74

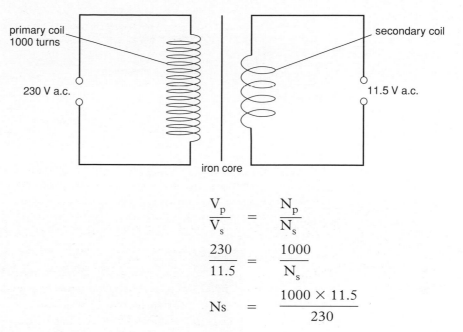

$$\frac{V_p}{V_s} = \frac{N_p}{N_s}$$

$$\frac{230}{11.5} = \frac{1000}{N_s}$$

$$N_s = \frac{1000 \times 11.5}{230}$$

Number of turns in secondary = 50 turns

Did you know?

Transformers can be constructed with very small energy losses. The large currents flowing in the coils produce unwanted heat energy. This is minimised by using low resistance windings in the coils. Large transformers have these windings specially cooled.

The changing magnetic fields linking the coils can produce eddy currents in the core material. These small electric currents would heat the core, so wasting energy. By building up the core with a large number of thin insulating sheets between the many magnetic layers, the eddy currents are greatly reduced.

Some transformers are almost 100% efficient, and many others approach this. The power transferred from the primary coils to the secondary coil stays almost constant.

Transmission of electricity

Power station generate electricity at voltages of 25 000 V (25 kV). This voltage is stepped up by a transformer to 400 000 V or 275 000 V. The electric power is transferred by cables in the transmission lines of the National Grid to all parts of the country.

A typical power stations produces 10 MW (1 megawatt = 1 000 000 W). If the power is transmitted at a low voltage then the current is high as P = VI. Much of the electrical energy is wasted as heat in the cables. The current is much lower when the voltage is stepped up and power losses in the cables are greatly reduced.

Figure 1.75
Simplified diagram of power transmission

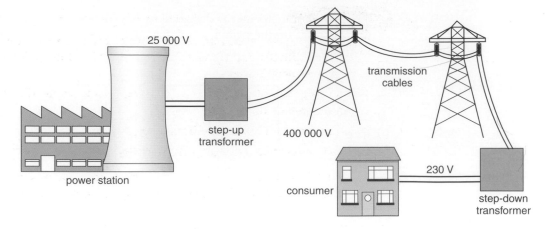

Figure 1.76
An electricity substation

At the consumer end, the voltage is stepped down to 230 V for houses. Every few streets in a town have an electricity substation, which is a step-down transformer serving the houses. The input voltage at these substations is very high and very dangerous. Most substations are open to the air to help them cool down.

The overhead transmission cables are usually made of aluminium because it has a low resistance. To reduce power losses in heating, the cables are thick. (Resistance is smaller for a thicker wire.) To reduce the weight and increase the rate of cooling, the cables are not covered by an insulation layer.

They are held far above the ground by ceramic insulating supports on pylons. Underground cables need costly insulation and maintenance work is also more difficult.

Did you know?

Aluminium transmission cables have a steel core. This is because aluminium is not strong enough to support the weight of the cable by itself.

Summary

- If a magnet is moved into or out from a coil of wire, or if a coil moves in a magnetic field a current is induced in the wire. This effect is called **electromagnetic induction**.

- The direction of the induced current depends on the direction of movement of the magnet or the coil.

- This effect is made use of in the a.c. generator.

- **Transformers** step-up or step-down voltages. They are used in the National Grid.

- Electrical power is transmitted at very high voltages. The small current in the power lines means less energy is wasted heating the cables and thinner cables can be used.

- Transformers consist of two coils wound on a single iron **core**.

- An alternating voltage in one coil induces an alternating voltage in the second coil.

- The voltages and the number of turns are related by the equation:

$$\frac{\text{voltage across primary}}{\text{voltage across secondary}} = \frac{\text{number of turns on primary}}{\text{number of turns on secondary}}$$

Topic questions

1 What would you notice if a wire, connected to a sensitive ammeter, is moved through a magnetic field?

2 A coil of wire with 20 turns spinning 60 times per minute between the poles of a magnet induces a current to flow in the coil. What three changes can be made to increase the size of the induced current?

3 The diagram shows a coil connected to a centre reading ammeter and a magnet about to be pushed into the coil.
What will happen to the meter needle when the magnet:
a) moves into the coil?
b) stays still in the coil?
c) moves out of the coil?

4 The diagram shows the percentage of original energy transferred in the production and distribution of electricity

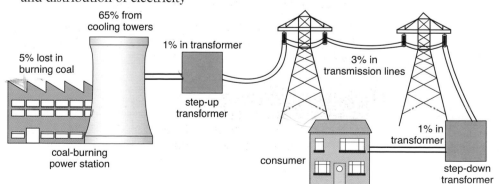

What percentage of the energy from the coal is
a) useful energy to the consumer?
b) lost in the transmission lines and transformers?
c) lost during production in the power station?

5 A simple transformer has 20 turns on the primary coil and 120 turns on the secondary coil. An alternating voltage of 12 V is supplied to the primary coil. What is the voltage across the secondary coil?

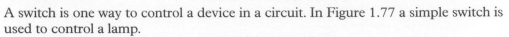

Control in circuits

Co-ordinated	Modular
	Mod 23
10.6	14.1 to 14.6

Switches

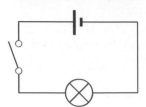

Figure 1.77

A switch is one way to control a device in a circuit. In Figure 1.77 a simple switch is used to control a lamp.

Closing the switch (the switch is ON) completes the circuit. A current will flow through the switch and the lamp will light.

Opening the switch (the switch is OFF) creates a break in the circuit. A current cannot flow through the switch so the lamp is off.

Not all switches are this simple. A **transistor** can be used as a high speed electronic switch (see pages 54–56).

Relays

Figure 1.78
Relay switches

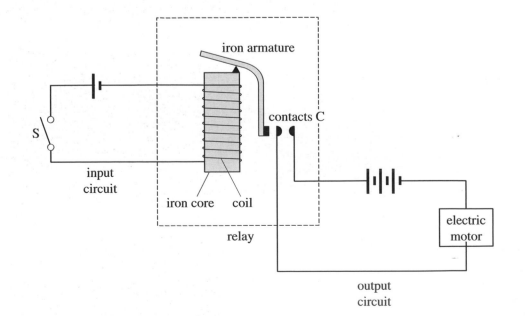

A **relay** is a switch worked by an electromagnet. A small current in the input circuit can be used to switch on a larger current in the separate output circuit. This makes the relay a useful switch to use in an electronic circuit (see page 51).

Figure 1.79

iron armature

contacts C

S

input circuit

iron core   coil

relay

electric motor

output circuit

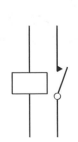

Figure 1.80 ▲
A circuit symbol for a normally open relay

Figure 1.79 shows how a relay can be used to switch on an electric motor. Closing the switch S completes the input circuit. A small current then flows through the coil, magnetizing the iron core. This attracts the iron armature, which as it pivots closes the contacts C. The output circuit is now complete and a current flows through the motor. This type of relay is called a 'normally open' relay. When the input switch is OFF (open), the contacts C are open, so the output circuit is OFF.

Robotic machines are usually computer controlled. The small current from the computer operates relays that switch on the larger currents needed to operate and control the robot.

Figure 1.81 ▶
A robot used to build cars

Resistors

The current through a circuit can be controlled using a fixed **resistor** or a variable resistor.

The fixed resistors used in electronic circuits often have coloured bands painted around them.

Figure 1.82
A circuit board containing resistors

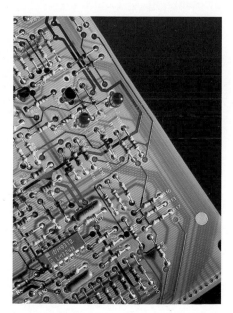

Electricity and Magnetism

The colours are part of a code used to indicate the value of the resistor and the tolerance (or accuracy) to which it is made.

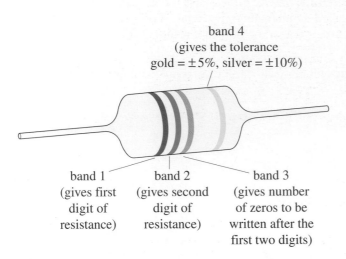

band 4
(gives the tolerance
gold = ±5%, silver = ±10%)

band 1
(gives first
digit of
resistance)

band 2
(gives second
digit of
resistance)

band 3
(gives number
of zeros to be
written after the
first two digits)

Figure 1.83
A resistor

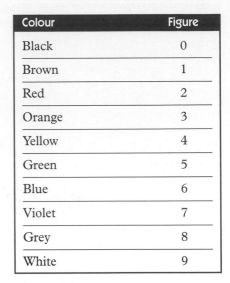

Colour	Figure
Black	0
Brown	1
Red	2
Orange	3
Yellow	4
Green	5
Blue	6
Violet	7
Grey	8
White	9

Figure 1.84
The resistor colour code

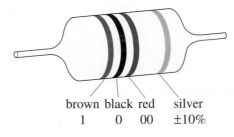

brown black red   silver
  1   0   00   ±10%

Figure 1.85
Resistance = 1000 ohms ± 10%
(note: 1000 ohms = 1 kilohm)

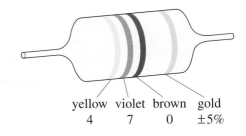

yellow violet brown gold
  4   7   0   ±5%

Figure 1.86
Resistance = 470 ohms ± 5%

A variable resistor can be used to vary the current in a circuit. The resistor consists of a long length of wire and a sliding contact. Moving the contact changes the resistance by changing the length of wire in the circuit.

Figure 1.87
A variable resistor

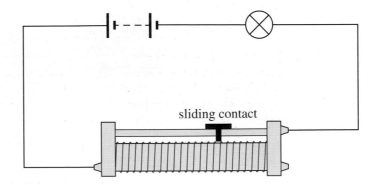

sliding contact

Moving the sliding contact to the left reduces the length of wire in the circuit. This reduces the resistance and increases the current. So the bulb gets brighter.

In radios, small variable resistors are used as volume controls. Turning the centre spindle changes the resistance and so changes the volume.

Figure 1.88
The volume control on a radio is a variable resistor

Topic questions

1 The diagram shows a simple circuit.

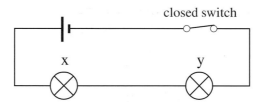

closed switch

x

y

What will happen to each lamp when the switch is opened? Give a reason for your answer.

2 The lamps in the circuit are identical.

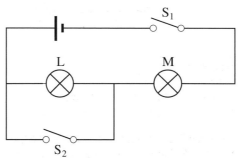

S_1

L

M

S_2

Explain what happens to each lamp when:
a) only switch S1 is closed;
b) only switch S2 is closed;
c) switch S1 and switch S2 are closed.

3 The circuit diagram shows a relay used to turn on an electric motor. Explain, step by step, why the motor turns on when switch S is closed.

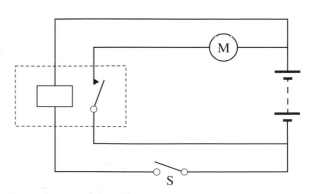

M

S

4   The diagram shows four resistors K, L, M and N. What is the value of each resistor?

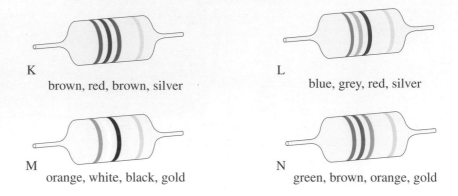

K   brown, red, brown, silver

L   blue, grey, red, silver

M   orange, white, black, gold

N   green, brown, orange, gold

5   What are the colours of the first three bands of the following value resistors?
a)   150 ohms
b)   11 ohms
c)   43 kilohms
d)   9.1 kilohms

6   A resistor has the following colour code:

orange   orange   orange   silver

What are the maximum and minimum values this resistor can have?

Electronic systems

Many electronic systems are designed in three parts, the **input sensors**, the **processor** and the **output device**. Systems can be shown as block diagrams. The arrows show the direction in which the signal or information is passed.

Input sensors detect changes in the environment. A sensor produces an electrical signal by transferring energy from one form (e.g. light) into electrical energy.

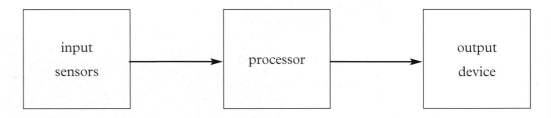

Input sensors include:

● Thermistors which detect changes in temperature.
● LDRs (light dependent resistors), which detect changes in light intensity.
● Microphones which respond to changes in sound intensity.
● Switches which respond to pressure, tilt, magnetic fields or moisture.

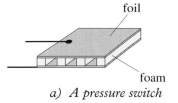

A simple pressure switch consists of two foil contacts separated by strips of foam. Applying pressure squashes the foam, making the contacts touch.

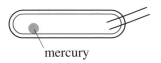

foil

foam

a) A pressure switch

Tilt switches close when they are held at an angle. The most common type uses a bead of mercury to close the gap between two contacts.

mercury

b) A tilt switch

Metal contacts are closed or opened by either a permanent magnet or an electromagnet. If the magnet closes the contacts it is a 'normally open' switch.

c) A magnetic switch

A moisture switch consists of two contacts close together. A current will flow when the gap between the contacts becomes moist.

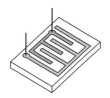

d) A moisture switch

Figure 1.89 *Types of input sensor*

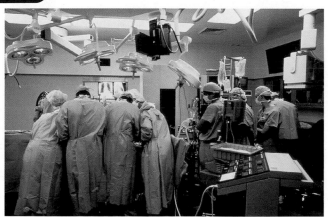

Figure 1.90

A hospital anaesthetist can use an infrared sensor to measure the carbon dioxide concentration in the air breathed out by a patient.

A processor takes the information from the input sensors and decides what action is needed. Processors can be made using logic gates (see page 50).

The output device is controlled by the processor. It transfers electrical energy supplied by the processor into other forms of energy.

Output devices include:

- Lamps which produce light.
- LEDs which produce light.
- Buzzers which produce sound.
- Electric motors which produce movement.
- Electric heaters which produce heat.
- Loudspeakers which produce sound.

Figure 1.92
LED circuit symbol

Figure 1.91
An LED

An LED (light emitting diode) is a special type of diode that glows when a small current passes through it. LEDs are often used as indicator lamps (on/off).

Logic Gates

Logic gates are switches used to control a wide range of devices. They are called gates because they only 'open' to produce an output signal for the right combination of input signals.

The circuits shown in Figures 1.93 and 1.95 are simple examples of logic gates. In each circuit the output from the lamp is determined by the inputs from the switches.

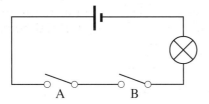

Figure 1.93
AND logic gate

Switch A	Switch B	Lamp output
open	open	OFF
open	closed	OFF
closed	open	OFF
closed	closed	ON

Figure 1.94
AND gate truth table

The lamp is ON when switch A AND switch B are closed.

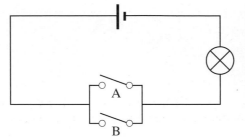

Figure 1.95
OR logic gate

Switch A	Switch B	Lamp output
open	open	OFF
open	closed	ON
closed	open	ON
closed	closed	ON

Figure 1.96
OR gate truth table

The lamp is ON when switch A OR switch B OR both switches are closed.

Figures 1.94 and 1.96 show the output from the lamp for every possible switch position. This sort of table is called a truth table.

In practice logic gates are made using combinations of transistors as the switches. This means they can be switched on and off millions of times a second.

AND, OR and NOT gates

In any circuit the inputs to a logic gate and the output from a logic gate can only be ON or OFF. An input or output that is ON is said to be HIGH. This is called logic state '1'. An input or output that is OFF is said to be LOW. This is called logic state '0'.

The three basic types of logic gate are called AND, OR and NOT. The circuit symbol and truth table for each of these gates is given in Figure 1.97.

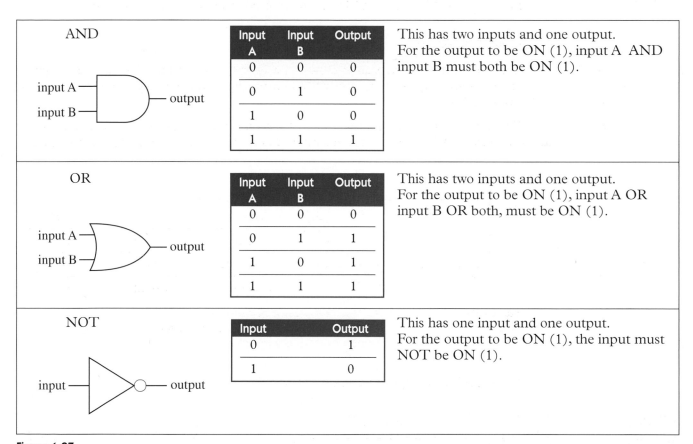

AND

input A ─┐
         ├─ output
input B ─┘

Input A	Input B	Output
0	0	0
0	1	0
1	0	0
1	1	1

This has two inputs and one output. For the output to be ON (1), input A AND input B must both be ON (1).

OR

input A ─┐
         ├─ output
input B ─┘

Input A	Input B	Output
0	0	0
0	1	1
1	0	1
1	1	1

This has two inputs and one output. For the output to be ON (1), input A OR input B OR both, must be ON (1).

NOT

input ─▷○─ output

Input	Output
0	1
1	0

This has one input and one output. For the output to be ON (1), the input must NOT be ON (1).

Figure 1.97
AND, OR and NOT gates

For many practical applications more than one logic gate is needed. The gates are sometimes combined so that the output from one becomes the input of another.

Examples:

● *Car door indicator*

Figure 1.98
A car door indicator

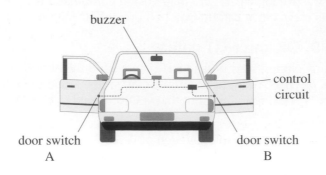

Switch A	Switch B	Buzzer
0	0	1
0	1	1
1	0	1
1	1	0

Figure 1.99
Truth table

Figure 1.100

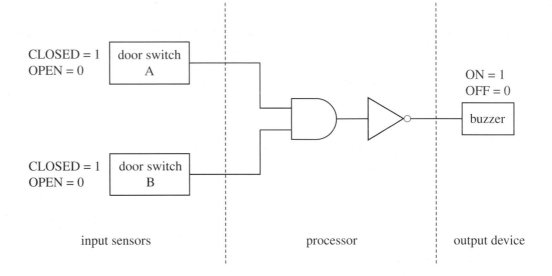

The buzzer is only OFF when both doors are closed.

● *Dawn to dusk security light*

Test switch	Light sensor	Security light
0	0	1
0	1	0
1	0	1
1	1	1

Figure 1.102
Truth table

Figure 1.101
Security light

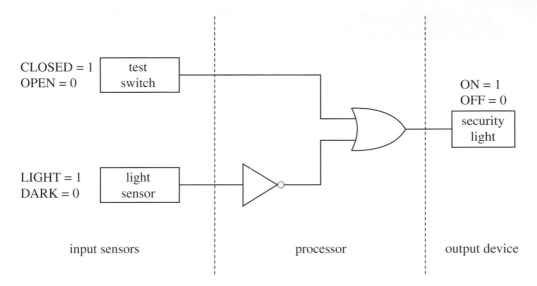

CLOSED = 1
OPEN = 0
test switch

LIGHT = 1
DARK = 0
light sensor

ON = 1
OFF = 0
security light

input sensors   processor   output device

Figure 1.103 ▲

The security light will automatically switch on when it is dark. The light can be tested at any time by simply closing the test switch.

- *Automatic heater for a greenhouse*

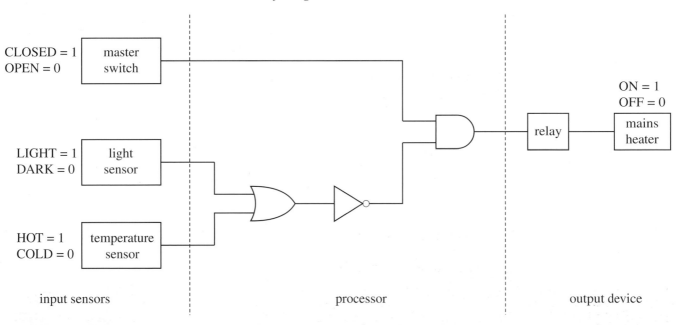

CLOSED = 1
OPEN = 0
master switch

LIGHT = 1
DARK = 0
light sensor

HOT = 1
COLD = 0
temperature sensor

ON = 1
OFF = 0
mains heater

relay

input sensors   processor   output device

Figure 1.104 ▲

Provided the master switch is closed, the heater will automatically switch on when it is both dark and cold. The heater can be switched off at any time by opening the master switch. The AND gate cannot provide enough power to operate the heater so it switches on a relay instead. The relay switches on the separate circuit of the heater (see page 42).

Figure 1.105
A greenhouse heater

Topic questions

1   What is the difference between an AND gate and an OR gate?

2   Which of the following can be used as an output device in an electronic system?

buzzer   heater   LDR   motor   switch   thermistor

3   Name two input sensors that could be used in a burglar alarm.

4

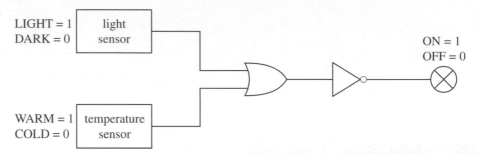

light	temperature sensor	lamp
0	0	
0	1	
1	0	

a)   Complete the truth table for the logic circuit above.
b)   Name a suitable device for use as the light sensor.
c)   Name a suitable device for use as the temperature sensor.
d)   What happens to the lamp when it is dark and cold ?
e)   What happens to the lamp when it is light and warm?
f)   Suggest a practical application for this circuit.

5   The diagram below shows part of the control system for an industrial guillotine. The guillotine only works when both switches are closed at the same time.

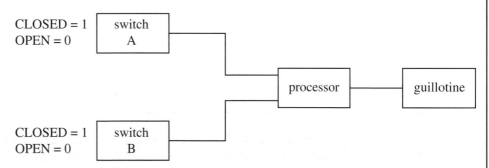

a)   Which part of the system is the output device?
b)   What type of logic gate should be used as the processor?

6   Draw the logic circuit for an alarm that will ring if either smoke or a high temperature is detected. Draw the truth table for your circuit.

Potential Divider

Two resistors joined in series can be used to split the voltage provided by a battery into two parts. Resistors used in this way form a **potential divider**. By changing the values of the resistors any fraction of the battery voltage can be obtained.

Figure 1.106
A potential divider

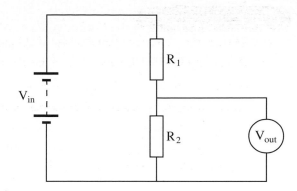

The output voltage of a potential divider can be worked out using the equation:

$$V_{out} = V_{in} \times \frac{(R_2)}{(R_1 + R_2)}$$

If either resistance R_1 or R_2 is increased (or reduced), the share of the input voltage across it also increases (or reduces).

Did you know?

The use of batteries has increased so much that more than 20 billion are sold each year.

Examples:

Figure 1.107

$$V_{out} = V_{in} \times \frac{(R_2)}{(R_1 + R_2)}$$

$$V_{out} = 9 \times \frac{2000}{(2000 + 2000)} = 4.5\text{V}$$

Figure 1.108

$$V_{out} = V_{in} \times \frac{(R_2)}{(R_1 + R_2)}$$

$$V_{out} = 9 \times \frac{4000}{(2000 + 4000)} = 6\text{V}$$

In control circuits an input sensor often replaces one of the resistors in the potential divider. Changes to the property that the sensor is detecting change the resistance of the sensor. This changes the value of the output voltage (V_{out}).

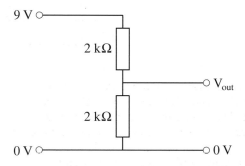

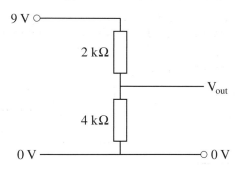

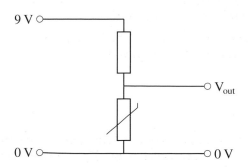

Figure 1.109

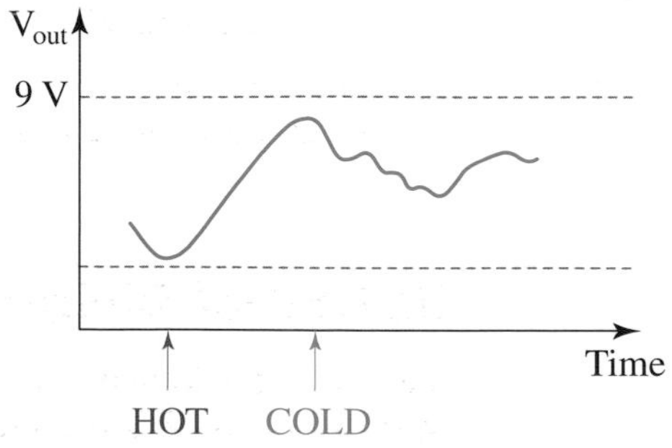

Figure 1.110
The effect of temperature on output voltage

The potential divider shown in Figure 1.109 has one fixed resistor and one thermistor. The value of the output voltage will change with temperature. When the temperature falls, the resistance of the thermistor goes up. So the share of the input voltage across the thermistor (V_{out}) goes up. V_{out} provides the input to the processor of the control circuit. If V_{out} changes enough, the input to the processor will change from being HIGH to being LOW (or from being LOW to being HIGH). In this particular circuit V_{out} could be used to switch on a heater.

It is often important that the input to the processor changes from HIGH to LOW at a particular value of the property that the sensor is detecting. For example, in a temperature control circuit a heater must switch on at a specific temperature. This can be done by using a variable resistor in place of the fixed resistor as in Figure 1.111.

Figure 1.111

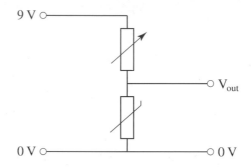

Increasing the resistance of the variable resistor causes the input to the processor to change from LOW to HIGH at a lower temperature.

Evaluating a circuit diagram

To explain how a control circuit works it is important to know what each component shown in the circuit diagram is doing.

As an example, consider the circuit shown in Figure 1.112. The circuit has been designed to work as a light dependent switch.

Figure 1.112

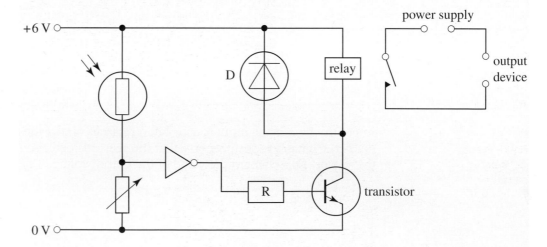

The LDR and variable resistor form a potential divider.

Since the resistance of the LDR changes with light intensity, the share of the input voltage across the LDR will change with light intensity. As it gets darker the resistance of the LDR and the voltage across the LDR go up. So the voltage across the variable resistor goes down.

At some point, the input to the NOT gate will change from HIGH to LOW, causing the output from the NOT gate to change from LOW to HIGH. A small current, limited by resistor R, then flows to the transistor, which switches ON. This allows a larger current to flow through the relay coil. The relay contacts close completing the output circuit.

Adjusting the variable resistor allows the output device to be activated at different light intensities. By decreasing the resistance, the input to the NOT gate changes from HIGH to LOW at a higher light intensity. So a brighter light than previously is able to activate the output device.

The function of diode D is to protect the transistor when the relay is switched off. Without the diode a large current would flow through the transistor, which might damage it.

Swapping the position of the variable resistor and LDR causes the output device to activate as the light intensity increases from dull to bright.

Replacing the LDR with a different type of sensor would allow the circuit to be used as a different type of switch. For example, by replacing the LDR with a moisture sensor, the circuit could be used to switch on an automatic water sprinkler system.

Adding a time delay to a control circuit

Sometimes a control circuit needs to include a time delay. For example, after setting a burglar alarm the system processor allows the operator several seconds to leave the building. The output device (siren) can only be activated after this time delay.

Figure 1.113
A microwave oven switches itself off after a set time delay

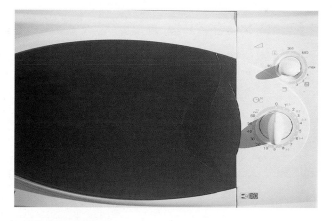

Including a **capacitor** in a circuit is one way to cause a time delay. A capacitor is a device designed to store electric charge.

Figure 1.114
Different types of capacitors

Figure 1.115
Capacitor circuit symbol

Electricity and Magnetism

Figure 1.116 shows a circuit used to charge up a capacitor. When the switch is closed a current flows to the capacitor and charge is stored by the capacitor. The potential difference (p.d.) across the capacitor, shown by the voltmeter, increases. The reading on the voltmeter will stop going up when the capacitor is fully charged. The reading on the ammeter will drop to zero when the capacitor is fully charged.

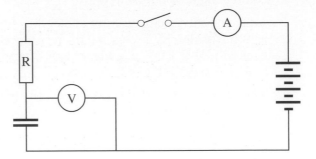

Figure 1.116

The larger the resistance of the circuit and the greater the value of the capacitor, the longer it takes for the capacitor to charge up.

Connecting a conductor across a charged capacitor causes the charge to flow away from the capacitor. The capacitor is said to be discharging. When a capacitor discharges, a current flows from the capacitor and the potential difference (p.d.) across the capacitor decreases. The capacitor is fully discharged when the p.d. across the capacitor is zero.

The greater the resistance of the discharging circuit and the greater the value of the capacitor, the longer it takes to discharge.

The rapid discharge of the capacitor in the flashgun of a camera produces a very large current for a very short time.

Figure 1.117
A camera flashgun

A time delay switch

Replacing an input sensor with a capacitor allows a time delay switch to be included in a control circuit.

Figure 1.118
A time delay switch

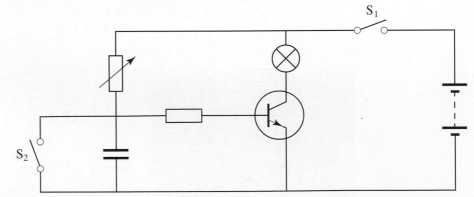

The capacitor and variable resistor form a potential divider. Closing switch S_1 (with S_2 open) causes the capacitor to charge up through the variable resistor. After a certain time the p.d. across the capacitor will have increased to the value needed to 'switch on' the transistor. Switching on the transistor completes the circuit to the lamp, so the lamp switches on.

The time delay between closing switch S$_1$ and the lamp lighting up can be changed by adjusting the value of the variable resistor or by changing the value of the capacitor. Increasing the value of the variable resistor or the capacitor would increase the time delay.

Closing switch S$_2$ discharges the capacitor and resets the timer.

Swapping the position of the variable resistor and capacitor would make the lamp come on the moment S$_1$ is closed. However as the voltage across the capacitor increases, the voltage across the variable resistor decreases. After a certain time, this causes the transistor to 'switch off' and the lamp to go out.

Advantages and disadvantages of advanced electronic systems

Advanced electronic systems are having an ever-increasing impact on everyday society. Since the invention of the transistor and microchip rapid technological developments have allowed electronic systems to become smaller, faster and more sophisticated. The way we work and the way we run our lives is continually changing as a result of these technological advances. However it should be realised that the advances made to electronic systems could have both advantages and disadvantages for society and the individual. Figure 1.119 outlines some of the possible advantages and disadvantages of four widely used electronic systems.

Figure 1.119
The advantages and disadvantages of electronic systems

Electronic system	Advantages	Disadvantages
Mobile phones	• Instant communication with almost anywhere in the world. • Ability to access information 'on the move'. • Increased mobility of workforce. • Ability to call for help in an emergency.	• Possible health hazard, linked to the use of microwave radiation. • Increased levels of crime. • Visual pollution and health worries, linked to the erection of phone masts. • Reduced levels of road safety as some car drivers phone whilst driving.
Closed Circuit Television (CCTV)	• Increased levels of security. • More crimes are detected and solved.	• Relocates crime to an area without CCTV. • Invasion of privacy, you may be filmed even though you are doing nothing wrong.

A CCTV Camera

Electronic system	Advantages	Disadvantages
Robotics *A bomb disposal robot*	• Can perform many simple, dull, routine or dangerous jobs. For example, car assembly, paint spraying or bomb disposal. • Reduces human error in manufacturing and improves reliability. For example in a clothes-manufacturing factory a robotic knife can cut through many layers of material to an exact pattern. • New job opportunities in the design, manufacture and maintenance of robotic systems.	• Higher levels of unemployment in the traditional manufacturing industries. One robot may replace a large number of factory workers.
Internet	• Access to huge amounts of information and data. • Ability to go shopping without leaving your home. • Worldwide communication. • Access to services, such as banking, at any time of the day or night. • A valuable teaching and educational aid.	• The possibility of transmitting unsuitable material to adults and/or children. • Invasion of privacy, due to the storage and transmission or personal information. • Organisations are vulnerable to computer 'hackers'.

Summary

◆ A switch is one way to control a device in a circuit.

◆ The current through a circuit can be controlled using a fixed resistor or a variable resistor.

◆ Most electronic systems are designed in three parts, the input sensors, a processor and an output device.

◆ The three basic types of logic gate are called AND, OR and NOT.

◆ A potential divider can be used to split the voltage provided by a battery into two parts.

◆ A capacitor can be used in a control circuit to cause a time delay.

◆ The advances made to electronic systems can have both advantages and disadvantages for society and the individual.

Topic questions

1 Consider the circuit shown in Figure 1.112 on page 54.
   a) What changes must be made if the circuit is to switch on a fan when the temperature in a room reaches 30°C?
   b) How can the fan be made to switch on at a lower temperature?

2 Adapt the circuit shown in Figure 1.118 on page 56 so that it can be used:
   a) as an electronic egg-timer for a blind person;
   b) to switch on a 230V lamp in a photographic dark room for a set time.

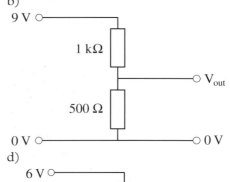

3 For each of the following circuits, calculate the value of V$_{out}$.

a)

9 V — 3 kΩ — V$_{out}$ — 6 kΩ — 0 V to 0 V

b)

9 V — 1 kΩ — V$_{out}$ — 500 Ω — 0 V to 0 V

c)

6 V — 18 Ω — V$_{out}$ — 12 Ω — 0 V to 0 V

d)

6 V — 5 kΩ — V$_{out}$ — 10 kΩ — 0 V to 0 V

4 A capacitor is discharged through a resistor. How can the discharge time be increased?

5 Outline the reasons why some people feel that privacy is no longer possible.

6 What types of job can a robot do well and what types of job does a human worker do better?

7 Some people think that mobile phones are dangerous. Why is this?

Examination questions

1 A student did an experiment with two strips of polythene. She held the strips together at one end. She rubbed down one strip with a dry cloth. Then she rubbed down the other strip with the dry cloth. Still holding the top ends together, she held up the strips.

a) i) What movement would you expect to see?
   (1 mark)
   ii) Why do the strips move in this way?
   (2 marks)

b) Copy and complete the **four** spaces in the passage.
   Each strip has a negative charge. The cloth is left with a _____ charge. This is because particles called _____ have been transferred from the _____ to the _____ .
   (4 marks)

2 a) The diagram shows a 13 amp plug.

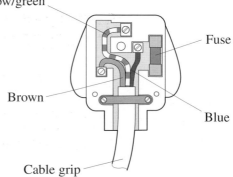

Yellow/green

Fuse

Brown

Blue

Cable grip

i) What is wrong with the way this plug has been wired? *(1 mark)*

ii) Why do plugs have a fuse? *(1 mark)*

b) The diagram shows an immersion heater which can be used to boil water in a mug.

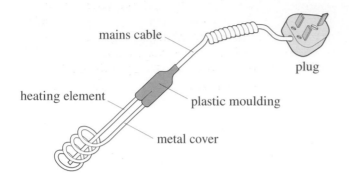

mains cable

plug

heating element

plastic moulding

metal cover

i) Which part of the immersion heater should be connected to the earth pin of the plug? *(1 mark)*

ii) Complete the sentence by choosing the correct words from the box. Each word may be used once or not at all.

| chemical | electrical | heat | light |

When the immersion heater is switched on _____ energy is transferred to _____ energy. *(2 marks)*

3 a) Look at this table of results.

VOLTAGE (V)	0.0	3.0	5.0	7.0	9.0	11.0
CURRENT (A)	0.0	1.0	1.4	1.7	1.9	2.1

i) Plot a graph of current agains voltage. Place current, in amps, on the vertical axis and voltage, in volts, on the horizontal axis. *(3 marks)*

ii) Use your graph to find the current when the voltage is 10 V. *(1 mark)*

iii) Use your answer to (ii) to calculate the resistance of the lamp when the voltage is 10 V. *(2 marks)*

b) i) What happens to the resistance of the lamp as the current through it increases?

ii) Explain you answer. *(2 marks)*

4 The drawing shows an experiment using a low voltage supply, a joulemeter, a small immersion heater and a container filled with water.

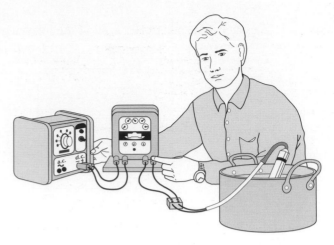

The potential difference was set at 6 V d.c. The reading on the joulemeter at the start of the experiment was 78 882 and 5 minutes later it was 80 142.

a) Use the equation:

$$\text{potential difference} = \frac{\text{energy transferred}}{\text{charge}}$$

to work out the total charge which flowed through the immersion heater in five minutes. Clearly show how you get to your answer and give the unit. *(3 marks)*

b) Calculate the current through the immersion heater during the 5 minutes. Write the equation you are going to use, show clearly how you get to your answer and give the unit. *(3 marks)*

5 The diagram shows a simple electricity generator. Rotating the loop of wire causes a current which lights the lamp.

State **three** ways to increase the current produced by the generator. *(3 marks)*

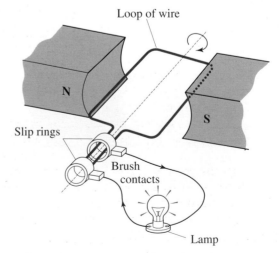

Loop of wire

N

S

Slip rings

Brush contacts

Lamp

6 A fault in an electrical circuit can cause too great a current to flow. Some circuits are switched off by a circuit breaker.

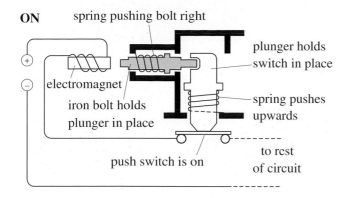

ON spring pushing bolt right

plunger holds switch in place

electromagnet

iron bolt holds plunger in place

spring pushes upwards

push switch is on

to rest of circuit

One type of circuit breaker is shown above. A normal current is flowing.

Explain, in full detail, what happens when a current which is bigger than normal flows.

(4 marks)

7 a) The diagram represents a simple transformer used to light a 12 V lamp. When the power supply is switched on the lamp is very dim.

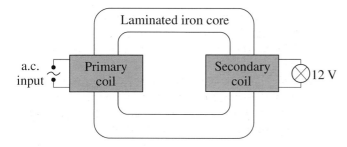

Laminated iron core

a.c. input

Primary coil

Secondary coil

12 V

Give **one** way to increase the voltage at the lamp with without changing the power supply.

(1 mark)

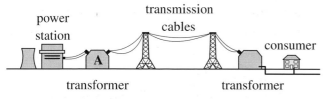

power station

transmission cables

consumer

transformer

transformer

b) Electrical energy is distributed around the country by a network of high voltage cables.

i) For the system to work the power is generated and distributed using alternating current rather than direct current. Why?

(1 mark)

ii) Transformers are an essential part of the distribution system. Explain why. *(2 marks)*

iii) The transmission cables are suspended high above the ground. Why? *(1 mark)*

c) The power station generates 100 MW of power at a voltage of 25 kV. Transformer **A**, which links the power station to the transmission cables, has 44 000 turns in its 275 kV secondary coil.

i) Write down the equation which links the number of turns in each transformer coil to the voltage across each transformer coil. *(1 mark)*

ii) Calculate the number of turns in the primary coil of transformer **A**. Show clearly how you work out your answer. *(2 marks)*

d) The diagram shows how the cost of transmitting the electricity along the cables depends upon the thickness of the cable.

Why does the cost due to the heating losses go down as the cable is made thicker?

(1 mark)

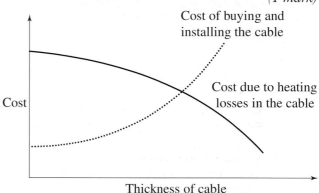

Cost of buying and installing the cable

Cost

Cost due to heating losses in the cable

Thickness of cable

8 The diagram below shows a circuit which can be used as an automatic switch.

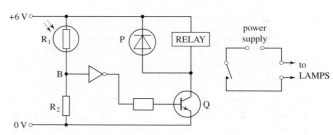

+6 V

R_1

P

RELAY

power supply

B

to LAMPS

R_2

Q

0 V

a) Name the following components: P, Q, R_1.

(3 marks)

Use the following information for parts b) and c).

$$V_{out} - V_{in} \times \frac{(R_2)}{(R_1 + R_2)}$$

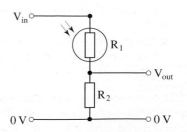

V_{in}

R_1

V_{out}

R_2

0 V

0 V

61

b) The resistance of R_2 = 2000Ω.
  V_{in} is 6V.
  i) In daylight the resistance of R_1 = 500Ω.
      Calculate the voltage across R_2.
  ii) In daylight the lamps will be OFF.
      Explain why. *(6 marks)*
c) In the dark the resistance of R_1 is 198000Ω.
   Calculate the voltage across R_2. *(2 marks)*

9 a) The diagram shows part of a simple alarm
  system used to protect a valuable necklace.

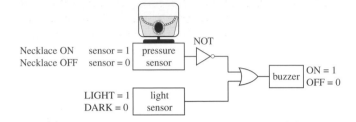

Necklace ON   sensor = 1
Necklace OFF sensor = 0

ON = 1
OFF = 0

LIGHT = 1
DARK = 0

  i) Copy and complete the truth table for the
      NOT gate.

Input	Output
1	
0	

(1 mark)

  ii) Copy and complete the truth table for the
      alarm system.

Pressure sensor	Light sensor	Buzzer
0	0	
0	1	
1	0	
1	1	

(2 marks)

  iii) Explain how this alarm system would
       work. *(2 marks)*
b) The alarm needs to be able to be switched on
   and off. To do this a key-operated switch and a
   logic gate **X** are added to the circuit.

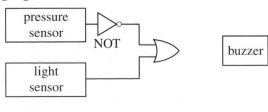

NOT

buzzer

ON = 1
OFF = 0

key-operated switch

X

  i) What type of logic gate is **X**? *(1 mark)*

  ii) Copy and complete the circuit above to
      show how the key-operated switch and
      logic gate **X** should be connected into the
      alarm system. *(2 marks)*

10 a) The diagram shows the arrangement of the
   colour coded bands on a typical resistor.

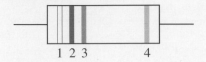

1 2 3   4

The colour code is given in the table below.

Figure	Colour
0	black
1	brown
2	red
3	orange
4	yellow
5	green
6	blue
7	violet
8	grey
9	white

  i) What are the colours of the first **three**
      bands of a 20 kΩ resistor? *(2 marks)*
  ii) What information is given by the
      fourth band? *(1 mark)*
b) The diagram shows two resistors joined in
   series. The variable resistor can have any
   value between 0 and 20 kΩ.

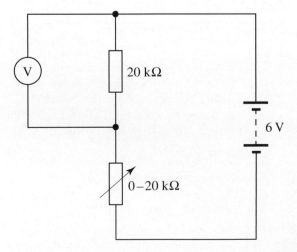

V

20 kΩ

6 V

0–20 kΩ

  i) What is the smallest possible reading on
      the voltmeter? *(1 mark)*

ii) What is the largest possible reading on the voltmeter? *(1 mark)*

c) The diagram shows one design for a time-delay circuit.

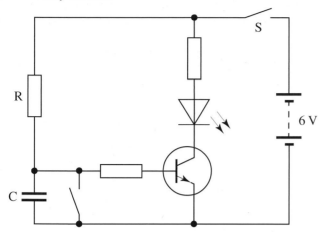

i) What is the function of a capacitor? *(1 mark)*

ii) When the switch **S** is closed, it is several minutes before the light emitting diode (LED) comes on. Explain why. The explanation has been started for you.

*When the switch **S** is closed, the voltage across the capacitor . . .* *(2 marks)*

iii) Give **one** practical use for this circuit. *(1 mark)*

iv) A pupil wires up the circuit. By mistake the positions of capactior **C** and the resistor **R** are swapped. Describe what will happen after the switch **S** is closed. *(2 marks)*

11 In the circuit shown below all four lamps are identical.
All four switches are closed (ON).

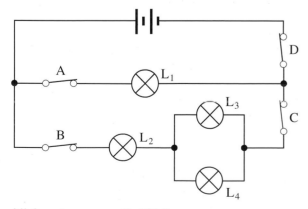

All four lamps are lit (ON).
a) Which **single** switch, **A** to **D**, should be opened in order to
 i) turn OFF **all four** lamps?
 ii) turn OFF one lamp only? *(2 marks)*

b) When **all four** switches are closed (ON), state which lamp L_1 to L_4 will be the brightest. Give a reason for your answer. *(2 marks)*

c) Lamps are sometimes used in electronic systems as output devices. Other devices are used as input sensors.
Below there is a list of output devices and input sensors.
Identify the **three** input sensors.

**buzzer   heater   LDR   motor
switch   thermistor**   *(3 marks)*

12 a) The diagram shows part of a heating system. It is designed to switch on automatically when it is both cold and dark. The control box contains two logic gates which are not shown.

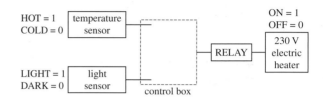

i) What is the name and circuit symbol for an input sensor which responds to light? *(2 marks)*

ii) What is the name and circuit symbol for an input sensor which responds to temperature? *(2 marks)*

iii) Copy and complete the truth table for the control system.

Light sensor	Temperature sensor	Heater
1	1	
1	0	
0	1	
0	0	

(2 marks)

iv) Identify the names of the **two** logic gates that should be used inside the control box, from the list below.

AND   NOT   OR   *(1 mark)*

v) Copy and complete the diagram in part a) to show how the two logic gates are used to connect the input sensors to the relay. Use the correct symbols for the logic gates. *(3 marks)*

vi) Why must a relay be used to operate the
    heater?   *(1 mark)*

b) The diagram shows an additional logic gate
    and switch added to the system.

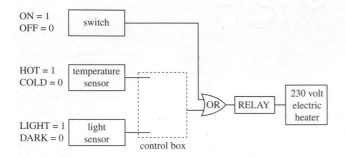

    Explain how this change allows the heater to be
    switched on at any time. The explanation has
    been stared for you.

    Closing the switch sends . . .   *(2 marks)*

Chapter 2
Forces and motion

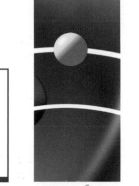

2.1	
Co-ordinated	**Modular**
	Mod 11
10.7	**12.1**

Speed, velocity and acceleration

If you travel at a fast **speed**, it takes less time to finish a journey than when you travel at a slow speed. Some passenger trains are so fast they can travel 280 kilometres in just one hour. In other words, the train has a speed of 280 kilometres per hour (280 km/h). This is the average speed of the train. During a one hour journey the train will sometimes go faster and sometimes slower.

Average speed can be worked out using this formula:

$$\text{average speed (in metres per second)} = \frac{\text{distance travelled (in metres)}}{\text{time taken (in seconds)}}$$

Example: In a race, a horse travels 1280 metres in 80 seconds. What is the average speed of the horse in metres per second?

distance travelled   =   1280 metres (m)

time taken   =   80 seconds (s)

$$\text{average speed} = \frac{\text{distance travelled}}{\text{time taken}} = \frac{1280\text{m}}{80\text{ s}} = 16 \text{ m/s}$$

Figure 2.1
Race horses in action

Forces and motion

On a journey, it's not just speed that is important, direction also counts. Figure 2.2 shows two routes which can be taken by a pupil going to school.

Figure 2.2

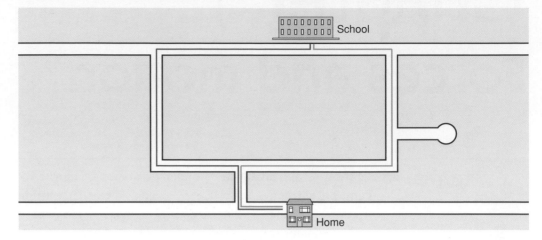

Both routes are the same distance and take the same time, so the speed is the same even though the directions are different. But each time the pupil changes direction, the **velocity** changes. This is because velocity is the speed of an object in a particular direction. If direction changes, velocity changes, even though the speed may stay the same. So to give a value for velocity, both speed and direction must be given.

Distance–time graphs

A distance–time graph can be used to work out the speed of an object.

Figure 2.3 shows that it takes 5 seconds to travel 30 metres. So the average speed can be calculated using:

$$\text{average speed} = \frac{\text{distance travelled}}{\text{time taken}} = \frac{30 \text{ m}}{5 \text{ s}} = 6 \text{ m/s}$$

This value is equal to the slope (or gradient) of the line of the graph. So, the slope of a distance–time graph equals the speed.

A distance–time graph can also be used to describe the movement of an object.

In Figure 2.4 the distance travelled is staying the same. This object is not moving – it is stationary. So, when an object is stationary, the distance–time graph is flat.

In Figure 2.5 the object is travelling the same distance each second. This object is described as moving at a constant speed. So, the steeper the slope of the straight line, the faster the speed.

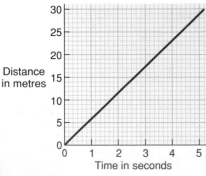

Figure 2.3

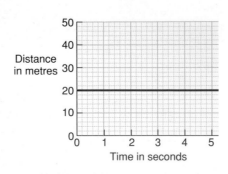

Figure 2.4

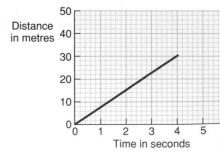

Figure 2.5

Figure 2.6

In Figure 2.6 the distance the object is travelling each second is increasing. This shows that the speed of the object is increasing. The object is said to be accelerating.

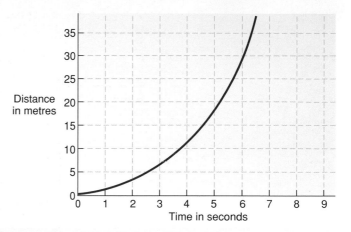

Using a distance-time graph

Figure 2.7 shows a distance–time graph for a cyclist. The slope of the line can be used to calculate the cyclist's speed.

Figure 2.7

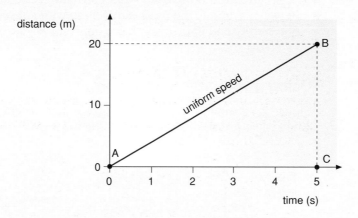

The slope of the line AB in Figure 2.7 $= \dfrac{B - C}{C - A}$

B – C = distance travelled (m)

C – A = time taken (s)

So, $\dfrac{B - C}{C - A} = \dfrac{\text{distance travelled}}{\text{time taken}} =$ speed (m/s)

In this example the slope of AB $= \dfrac{20}{5} = 4$, so the cyclist was moving at a steady speed of 4 m/s.

Velocity–time graphs

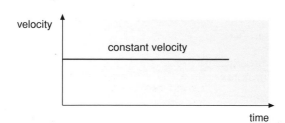

Figure 2.8

Figure 2.8 shows a velocity–time graph for a car travelling at a constant speed along a straight road. This means that the car is travelling with a constant velocity.

Figure 2.9

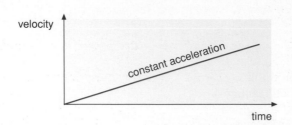

Figure 2.9 shows the velocity–time graph for a car accelerating along a straight road. The steeper the slope the greater the acceleration. A straight line for the slope means that the acceleration was constant.

Figure 2.10

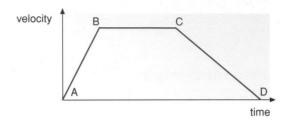

Figure 2.10 shows the velocity–time graph for a train moving from station A to station D along a straight track.

- Between A and B the train has a constant acceleration.
- Between B and C the train is travelling with constant velocity.
- Between C and D the train has a constant negative acceleration (**deceleration**).

Using a velocity–time graph

Figure 2.11 shows a velocity–time graph for a cyclist travelling in a straight line. The cyclist travels down a hill and then along a flat road.

Figure 2.11

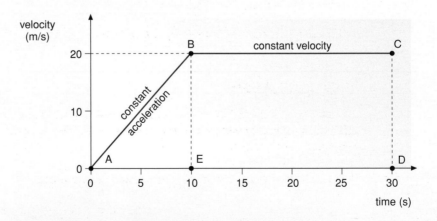

The slope of the line in a velocity–time graph can be used to calculate the acceleration of a moving object.

In Figure 2.11 the slope of line AB $= \dfrac{B - E}{E - A}$

B – E = change in velocity (m/s)

E – A = time taken (s)

So, $\dfrac{B - E}{E - A} = \dfrac{\text{change in velocity}}{\text{time taken}} = $ acceleration (m/s^2).

In this example:

slope AB $= \dfrac{B - E}{E - A} = \dfrac{20}{10} = 2$, so the cyclist is accelerating uniformly at 2 m/s^2

slope BC $= 0$, the cyclist is not accelerating but moving at a constant velocity.

The area under a line in a velocity–time graph can be used to calculate the distance travelled by the moving object. In Figure 2.11, the rectangular area under the line BC $=$ C $-$ B (or D $-$ E) $\times$ C $-$ D (or B $-$ E)

C $-$ B (or D $-$ E) $=$ time taken (s)
C $-$ D (or B $-$ E) $=$ velocity (m/s) $= \dfrac{\text{distance (m)}}{\text{time (s)}}$

So (C $-$ B) $\times$ (C $-$ D) $= \;\cancel{\text{time (s)}}\; \times \dfrac{\text{distance (m)}}{\cancel{\text{time (s)}}} =$ distance (m)

In this example the distance travelled by the cyclist while accelerating between A and B:

the area under line AB $= \frac{1}{2} \times$ (E $-$ A) $\times$ (B $-$ E)
$\qquad\qquad\qquad\qquad = 0.5 \times 10 \times 20$
$\qquad\qquad\qquad\qquad = 100$
So the cyclist has travelled 100 m.

The distance travelled by the cyclist while going at constant velocity between B and C:

the area under line BC $=$ (C $-$ B) $\times$ (C $-$ D)
$\qquad\qquad\qquad\qquad = 20 \times 20$
$\qquad\qquad\qquad\qquad = 400$
So the cyclist has travelled 400 m.

The total distance travelled by the cyclist is therefore 500 m.

Acceleration

When the velocity of an object changes, the object is accelerating. The faster the speed changes, the larger the **acceleration**. An object which is slowing down has a negative acceleration.

Acceleration can be worked out using this equation:

$$\text{acceleration (metres per second per second, m/s}^2) = \frac{\text{change in velocity (metres per second, m/s)}}{\text{time taken (seconds, s)}}$$

This can also be written as: $\quad a = \dfrac{v - u}{t}$

$$\text{acceleration} = \frac{\text{final velocity} - \text{starting velocity}}{\text{time taken (in seconds)}}$$

The unit of acceleration, metres per second per second, or metre/second squared, is usually written as m/s^2.

Example: At the start of a 100 metre race, an Olympic runner can accelerate to 12 metres per second in 2 seconds. What is the acceleration of the runner in m/s^2?

Figure 2.12 ▲
Olympic runners accelerating off the blocks

$$\text{starting velocity (u)} = 0 \text{ m/s}$$
$$\text{final velocity (v)} = 12 \text{ m/s}$$
$$\text{time taken (t)} = 2 \text{ s}$$

$$\text{acceleration} = \frac{\text{change in velocity}}{\text{time taken}} = \frac{12-0}{2} \text{ m/s}^2 = \frac{12}{2} \text{ m/s}^2 = 6 \text{ m/s}^2$$

Remember that velocity also involves direction. When the direction of an object changes, the velocity of the object changes. So whenever an object changes direction it has accelerated, even if the speed stays the same.

Summary

◆ Distance–time graphs can be used to show when an object is stationary or moving with a constant speed.

◆ The gradient of a distance–time graph represents the speed of an object.

◆ The velocity of an object is its speed in a particular direction.

◆ Velocity–time graphs can be used to show when an object is moving with constant velocity or with constant acceleration.

◆ The slope of a velocity–time graph can be used to calculate the acceleration of a moving object.

◆ The area under a velocity–time graph can be used to calculate the distance travelled.

◆ **Acceleration** is the rate at which the velocity of an object changes.

◆ Acceleration (m/s^2) = $\dfrac{\text{change in velocity (m/s)}}{\text{time taken (s)}}$

Topic questions

1 Copy and complete the following sentences.
   a) To work out speed you need to know the _____ travelled and the _____ taken.
   b) Velocity is the speed of an object in a particular _____ .

2 a) Copy and complete the following sentence by crossing out the two lines in the box which are wrong.

This graph shows an object which is moving at | zero / a constant / an increasing | speed.

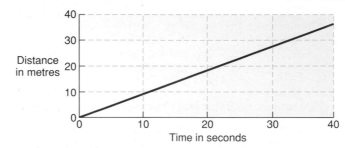

b) Calculate the average speed of this object.

3 For each of the animals, work out the missing quantity.

a) Cheetah: distance travelled = 75 m
              time taken = 2.5 s
              speed = ?
b) Antelope: distance travelled = 100 m
              time taken = 4 s
              speed = ?
c) Snail: time taken = 4000 s
              average speed = 0.0005 m/s
              distance travelled = ?
d) Swift: distance travelled = 200 m
              average speed = 50 m/s
              time taken = ?

4 A pupil cycles to school. The journey is shown on the distance–time graph below.

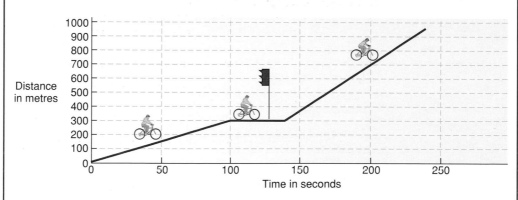

a) How far does the pupil live from school?
b) How long does it take for the pupil to cycle to school?
c) Work out the average speed of the pupil for the whole journey.
d) How do you know that the pupil had to stop at the traffic lights?
e) During which part of the journey was the pupil cycling the fastest?

5 A train starts from rest and accelerates uniformly along a straight track to reach a velocity of 30 m/s after 120 seconds. Draw a velocity–time graph for the train and use it to calculate:

a) the acceleration of the train
b) the distance travelled by the train.

71

2.2 Force and acceleration

A force has both size and direction. The size of a force is measured in **newtons** (N). In diagrams, a force is represented by an arrow. The longer the arrow, the larger the force.

If the forces are equal in size and opposite in direction, then the forces are balanced. Balanced forces do not change the velocity of an object. If an object is stationary, it will remain stationary. If an object is moving, it will continue to move at a constant speed in a straight line.

Figure 2.13 shows the forces acting on a flying aircraft.

Figure 2.13

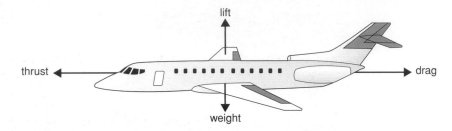

When the aircraft is flying at constant height and speed, lift is equal to **weight** and thrust is equal to **drag**. The vertical forces balance and the horizontal forces balance.

In a tug of war, two teams pull against each other. When both teams pull equally hard, the forces are balanced and the rope does not move. But when one team starts to pull with a larger force the rope moves. At this point the two forces are no longer balanced.

Figure 2.14

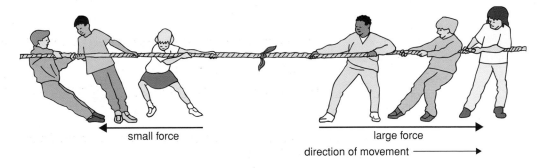

Unbalanced forces will change the velocity of an object. Since velocity involves both speed and direction, unbalanced forces can make an object speed up, slow down or change direction.

Unbalanced forces applied to the handlebars will make the cyclist change direction. This means the velocity of the cyclist will change even though the speed may stay the same.

Figure 2.15

An object will only accelerate when an unbalanced force acts on it. It then accelerates in the direction of the unbalanced force.

A car with a flat battery can usually be push started. With only one person pushing, the acceleration of the car is small, but the more people that push, the larger the acceleration. So, the larger the force the larger the acceleration.

Figure 2.16

Even four people would find it difficult to push start a van. This is because the **mass** of a van is far larger than the mass of a car. The larger the mass, the smaller the acceleration.

Figure 2.17

The force needed to accelerate a mass can be worked out using this equation:

force   =   mass   ×   acceleration
(newton, N)   (kilograms, kg)   (metre per second squared, m/s^2)

$$F = m \times a$$

Example: Calculate the force needed to give a train of mass 500 000 kg an acceleration of 0.5 m/s.

$$F = m \times a$$
$$\text{So, } F = 500\,000 \times 0.5$$
$$F = 250\,000 \text{ N}$$

This equation also shows that 1 newton is the force needed to give a mass of 1 kg an acceleration of 1 m/s^2.

Summary

◆ Balanced forces have no effect on the movement of an object. If it is stationary it will remain stationary, if it is moving it will carry on moving at the same speed and in the same direction.

◆ Unbalanced forces will affect the movement of an object.

◆ An unbalanced force on an object causes its velocity to change – it accelerates. The greater the force the greater the acceleration.

◆ The greater the mass of an object the greater the force needed to make it accelerate.

◆ One newton is the force needed to give a mass of one kilogram an acceleration of one metre per second squared.

◆ force (N) = mass (kg) × acceleration (m/s²)

Topic questions

1. Explain why a canoeist slows down when they stop paddling.

2. Why does a lorry need more powerful brakes than a car?

3. The diagram shows the four forces acting on a flying aircraft.
   a) When the plane is flying at constant speed, which two forces must be equal?
   b) When the plane is flying at a constant height above the ground, which two forces must be equal?

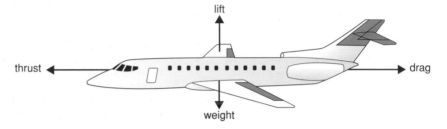

4. The diagram shows a ball which is about to be kicked.

Describe two different effects that the force of the kick will have on the ball.

5. The part of the space shuttle (orbiter) which returns to Earth has a mass of 78 000 kg and lands at a speed of 100 m/s. After touchdown it takes 50 s to decelerate and come to a halt.
   a) Calculate the deceleration of the orbiter.
   b) Calculate the force needed to bring the orbiter to a halt.

6. A cyclist travelling along a flat road stops pedalling. The speed drops from 8 m/s to 5 m/s in 6 s.
   a) Calculate the deceleration of the cyclist.
   b) The mass of the cyclist and cycle is 90 kg. Calculate the resistance force which slows the cyclist down.

Frictional forces and non-uniform motion

Co-ordinated	Modular
10.9	Mod 11 12.2

An engine produces the force needed to keep a car moving forwards. If the car is not accelerating, the force from the engine must be balanced by an equal force backwards. This force, called **air resistance** or drag, is a force of **friction**.

Figure 2.18

If the car engine stops, the car will slow down (decelerate). This happens because the force of friction opposes the motion of the car.

Friction also acts when one surface rubs against another. Without friction, the tyres of a car would not grip the road. The car would not be able to move forwards, backwards or turn.

Car brakes rely on friction. Pushing the brake pedal causes friction between the brakes and the wheels. The friction slows the wheels and stops the car. The total stopping distance of a car is made up of two parts: the **thinking distance** and the **braking distance**. The thinking distance is how far the car travels while the driver reacts to an emergency and applies the brakes. The braking distance is how far the car travels once the brakes have been applied. The force applied by the brakes will affect the braking distance. The larger the braking force, the shorter the braking distance.

Road conditions will also affect the braking distance. In icy or wet conditions the friction between the car tyres and the road is reduced. This reduces the grip and increases the braking distance. A rough surface will increase the friction and so reduce the braking distance.

Drivers' reactions are much slower if they have been drinking alcohol or if they are tired. The slower a driver's reaction time, the greater the thinking distance.

Speed affects both the thinking distance and the braking distance.

The faster the car, the greater the total stopping distance.

Figure 2.19

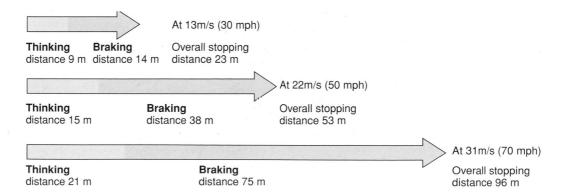

Force and non-uniform motion

If you want to know your weight, you will probably stand on some bathroom scales.

If the bathroom scales measure in kilograms, it does not show your weight, it shows your mass. Mass and weight are not the same thing. Mass is the amount of matter that makes up an object while weight is the force which **gravity** exerts on a mass. Mass is measured in kilograms (kg) while weight, like all forces, is measured in newtons (N). But more mass does mean more weight, because there is more mass for the force of gravity to pull on.

Figure 2.20

On Earth, gravity pulls on every one kilogram of mass with a force of 10 newtons. This is called the gravitational field strength (g).

$$g = 10 \text{ N/kg}$$

So someone with a mass of 50 kg will weigh

$$50 \times 10 = 500 \text{ N}$$

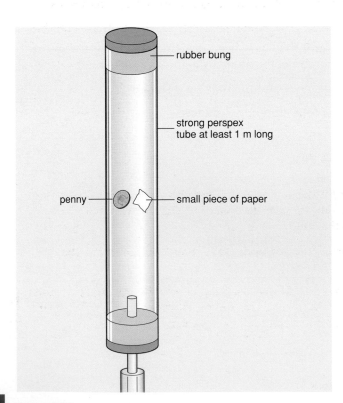

rubber bung

strong perspex
tube at least 1 m long

penny — small piece of paper

Gravity is a force of attraction which acts between objects. A ball thrown into the air will be attracted back towards the ground by gravity. As the ball falls downwards, gravity will cause it to accelerate. If there is no air resistance, the ball will accelerate at 10 m/s^2.

The acceleration due to gravity (g) = 10 m/s^2

This means that if there is no air resistance, the speed of a falling object will increase by 10 m/s every second. In a vacuum, where there is no air resistance, all falling objects accelerate at the same rate.

When the tube in Figure 2.21 is evacuated and then turned upside-down, the penny and piece of paper fall together. This is because the penny has more mass than the piece of paper. But this means gravity will exert a larger force on the penny, giving both objects the same acceleration.

Figure 2.21

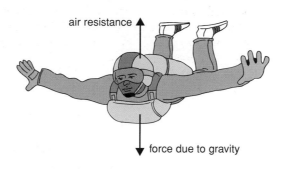

Usually air resistance does act on a falling object. Air resistance can only be ignored if the force it exerts is very small.

When sky-divers jump from a plane, the forces on them are unbalanced so the sky-divers accelerate.

Figure 2.22
Sky-diver at terminal velocity

air resistance ↑

↓ force due to gravity

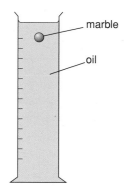

marble

oil

Figure 2.23

The faster the sky-diver falls, the larger the air resistance, so the smaller the acceleration. Eventually the downward force due to gravity and the upward force due to air resistance will be balanced. The sky-diver will stop accelerating and continue to fall at a constant speed. We say that the sky-diver has reached their **terminal velocity**.

Opening a parachute increases the air resistance. Since the force due to gravity stays the same, it will take less time before the two forces are balanced. This gives the sky-diver a lower terminal velocity.

Terminal velocity depends on the size of the air resistance force.

Air resistance is affected by the shape of an object. If a sky-diver curls up into a ball, air resistance will be reduced. It will take longer before the upward and downward forces balance, so the sky-diver reaches a higher terminal velocity. Terminal velocity therefore depends on the shape of an object.

Objects falling through liquids experience much larger resistance or drag forces.

A marble falling through oil will have a much lower terminal velocity than a marble falling through the air.

Did you know?

Vesna Vulovic, an air hostess, fell over 10 kilometres without a parachute and survived.

Summary

◆ The stopping distance of a moving vehicle depends on the braking force, the driver's reaction time, the condition of the road and the car, weather conditions, the speed and the mass of the car.

◆ For a vehicle travelling at a steady speed the frictional forces balance the driving force.

◆ Falling objects initially accelerate due to the force of gravity. As an object falls the air resistance increases until it balances the gravitational forces. When the resultant forces are balanced the object falls at its terminal velocity.

Topic questions

1 Explain the effect on the stopping distance of a car when there is ice on the road.

2 Why does a parachute slow down a falling parachutist?

3 Copy and complete the following sentence by crossing out in each box the two lines that are wrong.

When an astronaut goes into space his or her mass

will | increase
stay the same | and his or her
decrease

weight will | increase
stay the same | .
decrease

4 An astronaut is standing on the Moon and is about to let go of a hammer and a feather at the same instant.

What will happen, and why?

5 What is wrong with the following statement 'my weight is 80 kg'?

2.4   Turning forces

Co-ordinated	Modular
10.10	Mod 24
	15.1, 15.2

To undo a tight nut using just fingers is difficult. The job is made much easier if a spanner is used. The same force will then give a much larger turning effect. The longer the spanner, the larger the turning effect because the force will be further from the turning point or **pivot**.

Figure 2.24

pivot   force

The size of the turning effect about a pivot is called the **moment**. A moment can be worked out using the following equation:

$$\text{moment} = \text{force} \times \text{perpendicular distance between line}$$
moment (in newton-metres) = force (in newtons) × perpendicular distance between line of action and pivot (in metres)

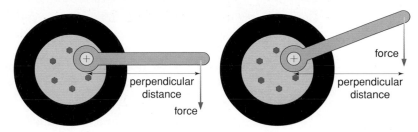

Moments are usually measured in newton-metres, this can be written as Nm. Figure 2.25 shows how the perpendicular distance between the line of action of the force and pivot is measured.

Figure 2.25
Measuring perpendicular distance

The person in Figure 2.26 is pushing the door as hard as possible, but it's not closing. This is because the person is pushing at the hinge and the hinge is the pivot. The distance between the force and the pivot is zero, so the moment is zero.

Example: Calculate the moment exerted by the cyclist in Figure 2.27 on the pedal.

$$\text{force} = 120 \text{ newtons (N)}$$
$$\text{distance} = 0.2 \text{ metres (m)}$$
$$\text{moment} = ?$$

Figure 2.26

moment = force × perpendicular distance between line of action and pivot

moment = 120 × 0.2 = 24 newton metres (Nm)

Figure 2.27

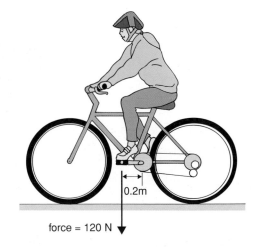

Figure 2.28 shows two children on a playground see-saw.

Figure 2.28

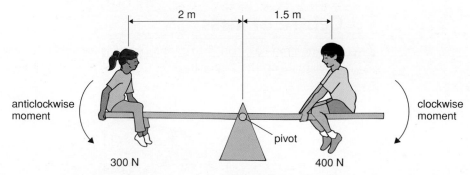

The moment of the boy is trying to turn the see-saw clockwise. The moment of the girl is trying to turn the see-saw anticlockwise. But the see-saw is not turning – it is balanced (in equilibrium). This means that the size of the clockwise moment must be the same as the size of the anticlockwise moment.

Figure 2.29 ▲

When an object is not turning:

total clockwise moment = total anticlockwise moment

Lots of joints in the human body act as pivots. Your elbow joint acts as a pivot, it lets your lower arm rotate up towards your shoulder and down towards your leg. Holding your lower arm horizontal will soon make your bicep muscle ache. This is because the clockwise moment produced by the weight of your lower arm must be balanced by an anticlockwise moment from the biceps muscle. Figure 2.30 shows the forces acting on your lower arm when you hold it horizontal.

$$\text{clockwise moment produced by the lower arm} = 15 \times 0.20$$
$$= 3 \text{ Nm}$$

$$\text{anticlockwise moment produced by the biceps muscle} = F \times 0.05$$

$$\text{total clockwise moment} = \text{total anticlockwise moment}$$

$$3 = F \times 0.05$$

$$F = \text{force in the} = 60 \text{ N}$$
$$\text{biceps muscle}$$

So the force exerted by the muscle to hold the lower arm horizontal is four times larger than the weight of the lower arm itself. This is because the muscle is attached so close to the pivot.

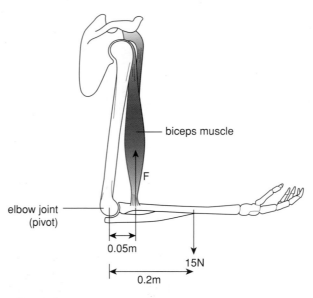

Figure 2.30 ▶

Centre of mass

Balancing a ruler on the tip of a finger is easy. But it only works if the centre of the ruler rests on the finger tip.

Figure 2.31

This balance point on the ruler is called the **centre of mass**. Although every part of the ruler has a mass, the centre of mass is the point where all the mass of the ruler can be thought to be concentrated. This is also the point through which the weight of the ruler acts (the centre of gravity). So if the finger tip (the pivot) is positioned at the centre of mass, the weight of the ruler will produce no clockwise or anticlockwise moment, the ruler is therefore balanced.

A tightrope walker will stay balanced if he keeps his centre of mass directly above the rope. The long pole helps the tight rope walker to keep his balance. If he starts to topple to one side of the rope, moving the pole to the other side will bring the centre of mass of the walker and the pole back over the rope.

Figure 2.32
A tightrope walker

The ruler in Figure 2.31 is symmetrical. The centre of mass of the ruler is at the centre of the ruler, along an axis of symmetry.

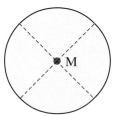

Figure 2.33 shows three common symmetrical shapes. The dotted lines are axes of symmetry.

For all symmetrical objects, the centre of mass (M) is along an axis of symmetry.

Figure 2.33 ▲

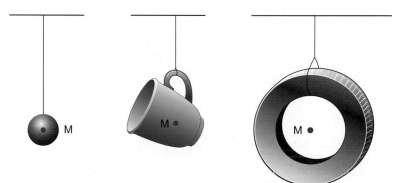

Figure 2.34 ▶

When an object is suspended it will come to rest with its centre of mass directly below the point of suspension. This is why a plumb-line always hangs vertically.

The centre of mass of an object is not always within the object itself. The centre of mass of the tyre in Figure 2.34 is in the air at the centre of the circle. But the tyre is still balanced.

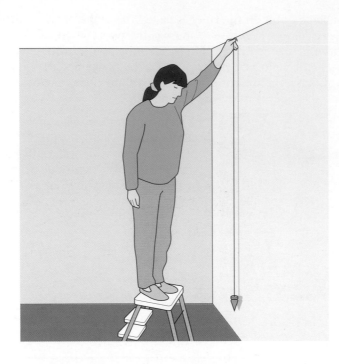

Figure 2.35
A decorator using a plumb line

By arching their back pole-vaulters can move their centres of mass outside their body. This lets them jump higher. The energy they use to lift themselves off the ground raises their centre of mass which passes under the bar. But their body passes over the bar.

Figure 2.36
Sergey Bubka in action

Finding a centre of mass

The centre of mass of an irregularly shaped piece of card can be found using a plumb-line. First a small hole is made in the card. The card is then suspended from a long pin. The card must be able to swing freely. A plumb-line is also suspended from the pin. When the card stops swinging, a line showing the position of the plumb-line is drawn on the card. The centre of mass of the card must be somewhere along this line. The card is now suspended from a different point. The plumb-line is used to draw a second vertical line. Where the two lines cross is the centre of mass of the card.

Figure 2.37
Finding the centre of mass

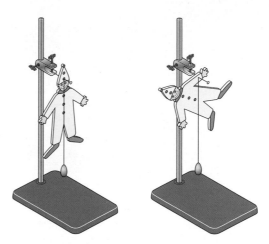

If the position of the centre of mass is accurate the card will balance at this point on the tip of a finger.

Figure 2.38

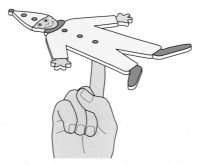

An object with a pivot not passing through its centre of mass will always try to turn. But the object can be balanced using an additional force. Imagine a window cleaner walking along carrying a ladder on his shoulder. A bucket of water hangs from one end of the ladder. The weight of the ladder (acting at the centre of mass of the ladder) causes an anticlockwise moment. But the ladder and bucket of water are balanced so there must be an equal clockwise moment. The weight of the bucket of water causes the clockwise moment.

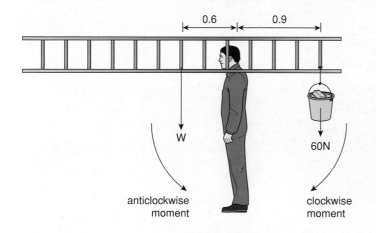

clockwise moment = 60×0.9

$= 54$ Nm

anticlockwise moment = $W \times 0.6$

clockwise moment = anticlockwise moment

$54 = W \times 0.6$

$W = 90$ N

Figure 2.39

Stability

Most objects are designed to be stable. This means that if they are tilted slightly and then released they will not fall over. They will fall back to their original position. If tilted too much the object will become unstable and fall over. So how much can an object be tilted and still be stable. Figure 2.40 shows two different table lamps. The vertical lines drawn from the centres of mass show the line of action of the weight of each lamp.

Figure 2.40

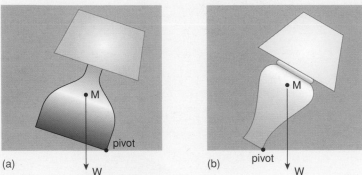

(a)

(b)

In Figure 2.40a, the line of action of the weight falls inside the base of the lamp. The lamp will fall back to its original position, it is stable. In Figure 2.40b, the line of action of the weight falls outside the base of the lamp. The turning effect now caused by the weight makes the lamp topple over. The lamp is unstable.

Figure 2.42
A tractor ploughing a hillside

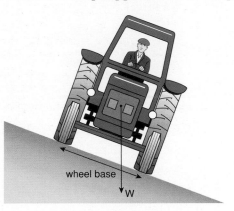

Figure 2.41

Tractors are often used on sloping ground, so stability is an important part of their design. Stability is achieved by having a low centre of mass and a wide wheel base. Figure 2.41 shows that even on sloping ground the line of action of the weight falls within the wheel base of the tractor, so the tractor is stable.

Did you know?

Speed skiers can exceed 160 km/hr. At this speed stability is very important.

Figure 2.43
A speed skier in action

Summary

◆ A moment can be worked out using the equation:

moment = force × perpendicular distance between line of action and pivot

◆ When an object is not turning, the total clockwise moment equals the total anticlockwise moment.

◆ The centre of mass is the point where all the mass of an object can be thought to be concentrated.

◆ If the line of action of the weight of an object lies outside the base of the object, the object will tend to fall over, it is unstable.

Topic questions

1 Copy and complete the following sentences:
   a) The turning effect of a force is also called its _____ .
   b) A suspended object will come to rest with its _____ of _____ directly below the point of _____.
   c) When an object is balanced, the total _____ moments must be _____ to the total _____ moments.

2 A person is trying to lever a nail out of a block of wood using a claw hammer. Why will the nail come out more easily if a hammer with a longer shaft is used?

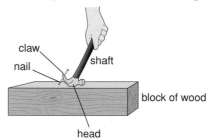

3 The diagram shows a spanner being used to undo a tight nut. Calculate the size of the turning moment exerted by the force on the nut.

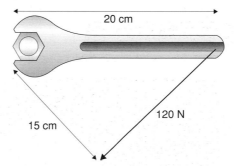

4 The diagram shows two children playing on a see-saw. If the see-saw is balanced what is the value of X?

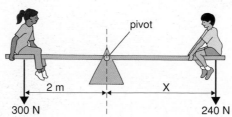

5 A crane is used to lift a container off a ship. Calculate the weight of the counterbalance needed to stop the crane from falling over.

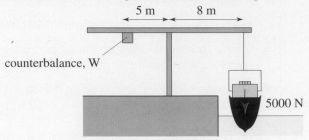

6 A builder needs to weigh a symmetrical plank of wood. The maximum range of the builders spring balance is less than the weight of the plank. The diagram shows how the builder overcame this problem.

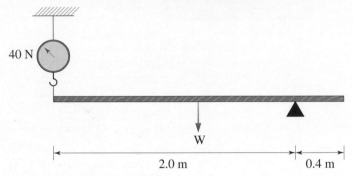

Calculate the weight of the plank.

7 The diagram shows a student leaning back on her stool. The stool and student are about to topple over.

a) Which one of the dots, A, B or C marks the centre of mass of the stool and student? Explain the reason for your choice.
b) How could the stool have been designed to make it more stable?

8 The diagram shows a simple balancing toy.

Given a small push the toy will rock backwards and forwards without falling off the bar. Explain why.

2.5 Momentum

A moving object has both **kinetic energy** and **momentum**. The greater the mass of the object and the faster it moves, the more kinetic energy and momentum it has. In some situations it is more useful to consider the momentum of an object rather than its kinetic energy.

The momentum of an object is defined by the equation:

momentum   =   mass   ×   velocity
(in kilogram metre/second)   (in kilograms)   (in metres/second)

The unit of momentum, kilogram metre/second, is usually written as kg m/s.

Just like **velocity**, momentum has both a magnitude (size) and a direction. The momentum of an object is always in the same direction as the velocity of the object. So if two objects move in opposite directions one will have positive momentum and the other negative momentum.

Figure 2.44
A charging rhino has lots of momentum

Example: A lorry and a car are travelling in opposite directions. Calculate the momentum of each vehicle.

Figure 2.45

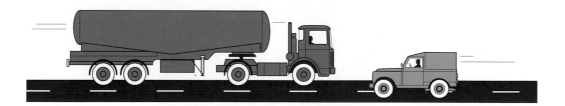

Lorry		Car	
mass	= 5500 kg	mass	= 800 kg
velocity	= 14 m/s to the right	velocity	= 25 m/s to the left
momentum	= mass × velocity	momentum	= mass × velocity
	= 5500 × 14		= 800 × 25
	= 77 000 kg m/s		= 20 000 kg m/s
momentum of the lorry	= + 77 000 kg m/s	momentum of the car	= − 20 000 kg m/s

The momentum of the lorry moving to the right has been taken to be positive, so the momentum of the car moving to the left must be given as negative.

Force and momentum

When a resultant force acts on an object, the object will accelerate. This means that the velocity, and therefore the momentum, of the object will change. So a resultant force acting on an object will cause a change in momentum.

The size of the resultant force and the change in momentum that occurs are linked by the following equation:

$$\text{force (in newtons)} = \frac{\text{change in momentum (in kilogram metre/second)}}{\text{time (in seconds)}}$$

A small force acting for a long time can cause the same change in momentum as a large force acting for a small time.

Golf and cricket are just two of the sports that demonstrate momentum.

Figure 2.46
A golfer follows through to give the ball more momentum

The golfer swings the club, hits the ball and follows through. The follow through means that the club does not stop the moment it hits the ball. The club continues, for a short time, to move with the ball. This action increases the time that the force acts on the ball, which increases the change in momentum of the ball. So the ball will move faster and travel further.

Figure 2.47
A cricketer draws his hands back to reduce the impact of the ball

Catching a fast moving cricket ball can be a painful experience. Using a large force to stop the ball quickly will produce an equally large force on the catcher's hands.

By pulling his hands backwards as he catches the ball, the cricketer takes longer to reduce the momentum of the ball to zero. This reduces both the force needed to stop the ball and the force exerted on the cricketer's hands.

Figure 2.48

Example: The picture shows a golfer about to strike a stationary golf ball of mass 0.045 kilogrammes.

When the golf club strikes the ball it is in contact for 0.001 seconds and exerts a force of 3600 newtons on the ball. Calculate the velocity at which the ball leaves the club.

$$\text{change in momentum} = \text{force} \times \text{time}$$

$$= 3600 \times 0.001$$

$$= 3.6 \text{ kg m/s}$$

since the ball was initially stationary,

$$\text{change in momentum} = \text{mass} \times \text{final velocity}$$

$$\text{So, final velocity} = \frac{\text{change in momentum}}{\text{mass}} = \frac{3.6}{0.045} = 80 \text{ m/s}$$

Momentum and safety

When an object collides with another, the two objects exert a force on each other. These forces are equal in size, but opposite in direction. The change in momentum of each object will be equal in size, but opposite in direction. The longer the time of contact the smaller the force needed to change the momentum. This is the principle behind many different types of safety devices.

Figure 2.49

A seat belt is not rigid. In a crash it is designed to stretch slightly. This is very important, since it increases the time taken for the momentum of the passenger to be reduced to zero. So both the force on the passenger's body and the subsequent risk of injury are reduced.

Figure 2.50

The helmet worn by a cyclist and the body protector worn by a horse rider work in the same way. If the cyclist falls off and hits his/her head, the padding inside the helmet will start to crush.

Figure 2.51

If the horse rider falls off, the padding inside the body protector will start to crush. In both cases the time taken to stop has been increased so reducing the force on the rider.

Collisions and explosions

In any collision or explosion, the momentum after the collision or explosion in a particular direction, is the same as the momentum in that direction before the collision or explosion.

That is:

$$\text{total momentum after a collision or explosion} = \text{total momentum before a collision or explosion}$$

Momentum is conserved provided there are no external forces acting.

Collisions

Consider the white snooker ball in Figure 2.52 moving with velocity 'v' directly towards a red snooker ball. Before they collide the red ball is not moving. The red ball has zero momentum.

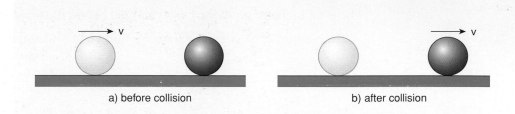

Figure 2.52

a) before collision   b) after collision

During the collision, an equal size force will act on each ball, but in opposite directions. The forces will also act for the same time. The momentum of each ball will therefore change by the same amount.

(*Remember*: force × time = change in momentum)

After the collision the white ball will be stationary. The white ball has zero momentum. The red ball will move with a velocity 'v' in the same direction as the white ball was moving. The total momentum after the collision is the same as the total momentum before the collision.

Momentum has been conserved, but what of kinetic energy?

(*Remember*: kinetic energy = ½ × mass × velocity²)

Before the collision only the white ball has kinetic energy. After the collision only the red ball has kinetic energy. The velocities before and after the collision are the same. So the total kinetic energy before and after the collision is the same. This is an **elastic collision**. Elastic collisions are those that involve no overall change in kinetic energy.

Most collisions are not perfectly elastic. Usually when objects collide, the total kinetic energy after the collision is less than before the collision. Some of the kinetic energy is transferred to other forms of energy.

Explosions

An explosion is the opposite of a collision. Instead of moving together, objects move apart. But like a collision, the total momentum of the objects involved in an explosion remains constant. In an explosion momentum is conserved.

Example: At the 'Circus on Ice' a clown throws a 'custard pie' forwards with as much force as possible. Before throwing the pie, the clown was not moving.

Figure 2.53

v = 10 m/s
mass pie = 0.75 kg

mass of clown
= 50 kg

Calculate the speed at which the clown moves backwards.

The total momentum of the clown and pie before throwing = 0

Momentum of clown and pie before throwing = Momentum of clown and pie
after throwing

0 = Forward momentum + backward momentum
of the pie of the clown

$0 = (0.75 \times 10) + (50 \times v)$

$0 = 7.5 + (50 \times v)$

$v = -0.15$ m/s

The negative sign means that the clown does move in the opposite direction to the pie.

Figure 2.54
The launch of a space shuttle

The principle of momentum conservation is used in the launch of a space shuttle. Before the launch the total momentum of the shuttle and fuel is zero. After lift off the total momentum must still be zero. So the downward momentum of the hot exhaust gases must be equal to the upward momentum of the shuttle.

Summary

◆ Momentum is defined by the equation:

momentum = mass × velocity

◆ The unit of momentum is the kilogram metre/second (kg m/s).

◆ Momentum has both magnitude (size) and direction.

◆ $\text{force} = \dfrac{\text{change in momentum}}{\text{time}}$

◆ total momentum after a = total momentum before
collision or explosion a collision or explosion
(provided no external forces act)

◆ Elastic collisions are those that involve no overall change in kinetic energy.

Topic questions

1 Copy and complete the following sentences:
   a) An object that is not moving has _____ momentum.
   b) Momentum, like velocity has both _____ and _____ .
   c) Momentum is worked out by multiplying _____ and
      _____ .

2 Work out the momentum of:
   a) a jogger of mass 75 kg running at 3 m/s.
   b) a cyclist of mass 60 kg peddling at 8 m/s
   c) a hockey ball of mass 0.5 kg moving at 12 m/s
   d) a bullet of mass 10 g moving at 250 m/s

3 A racing car, travelling at 85 m/s, has a momentum of 42 500 kg m/s. What is the mass of the racing car?

4 An elephant of mass 1800 kg has a momentum of 10 800 kg m/s. How fast is the elephant moving?

5 During a crash test, a car of mass 840 kg is driven into a wall at 5 m/s. The car slows down and stops in 1.5 seconds. Calculate the force on the car during the collision.

6 A trolley of mass 2 kg, travelling at 3 m/s, collides with a stationary 1 kg trolley. If the trolleys stick together, calculate:
   a) the total momentum before the collision,
   b) the speed of the two trolleys after the collision.

7 Explain how the speed of a rocket travelling through space can be increased.

8 A bullet is fired into a block of wood. The collision is not elastic. Explain why.

2.6	
Co-ordinated	**Modular**
10.12	**Mod 24** 15.3

Circular motion

Figure 2.55
A skydiver falls with a constant velocity

Figure 2.55 shows the forces acting on a skydiver. The two forces are equal in size and opposite in direction. So the forces are balanced.

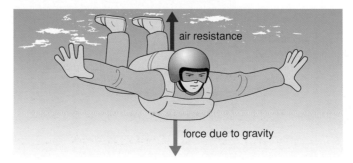

Balanced forces do not change the velocity of an object, so the skydiver will fall at a constant speed in a straight line. But objects do not always move in straight lines, sometimes they follow circular paths. This means that they are continually changing direction. But when the direction changes, the velocity of the object changes. For this to happen a resultant, unbalanced force is needed.

Forces and motion

Figure 2.56

A ball can be tied to a piece of string and whirled around in a horizontal circle.

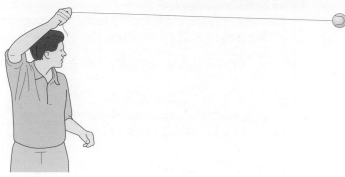

The speed of the ball stays the same but its direction is constantly changing. So its velocity is also changing. The force that causes the change in velocity is the inward pull of the string on the ball. This force is always towards the centre of the circle, at right angles to the direction the ball is travelling in. The force does not increase or decrease the speed of the ball. If the string breaks the ball would fly off in a straight line in the direction of travel, as in Figure 2.57.

Figure 2.57

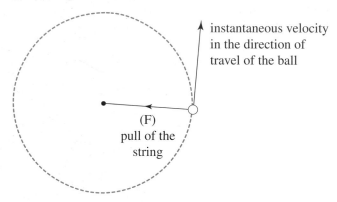

instantaneous velocity in the direction of travel of the ball

(F) pull of the string

The same idea is used in the sport of hammer throwing. The athlete makes the hammer move in a circle by using the pull of their arms. At the right moment the athlete lets go and the hammer flies off into the air.

Figure 2.58
A hammer thrower

Did you know?

The men's world hammer throwing record of 86.7 metres was set by Yuriy Sedykh in 1986.

The resultant force that makes an object move in a circle is called the **centripetal force**. This force always acts inwards, towards the centre of the circle. Although a centripetal force can be provided in different ways, the size of the centripetal force always depends upon the same three factors. The centripetal force is greater:

● the greater the mass of the object;
● the greater the speed of the object;
● the smaller the radius of the circle.

The speed of the object has the greatest effect on the size of the centripetal force. If speed doubles, the centripetal force increases four times.

Centripetal force in action

There must be a centripetal force on a motorbike when it goes around a bend in the road. This force is provided by the friction between the road and the motorbike's tyres. Ice on the road will reduce the friction. A motorcyclist must reduce speed if they want to stay in control and not skid off the road.

Figure 2.59
A motorcyclist taking a corner at high speed

Figure 2.60
A satellite orbiting the Earth

Many satellites move in circular orbits around the Earth. The centripetal force is provided by the gravitational pull of the Earth on the satellite. In a similar way the Earth is kept in orbit around the Sun by a centripetal force. In this case the force is provided by the gravitational pull of the Sun on the Earth.

Did you know?

Satellites orbiting close to the Earth have speeds of about 29 000 km/hr.

Figure 2.61

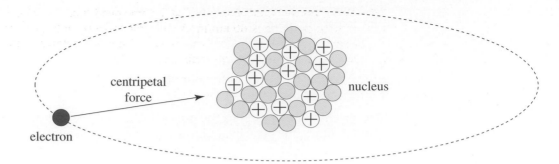

For electrons to orbit the nucleus of an atom there must be a centripetal force. The force is provided by the electrostatic attraction between the negative electron and the positive nucleus.

Summary

◆ Any object moving in a circle is acted on by a centripetal force.

◆ The centripetal force on an object is affected by the mass of the object, the speed of the object and the radius of the object's circular path.

Topic questions

1 Copy and complete the following sentences.
   a) The _____ force needed when a car travels around a bend in the road is provided by the force of _____ between the road and the car _____ .
   b) The force keeping the Moon in orbit around the Earth is the _____ pull of the _____ on the Moon.
   c) The faster a bicycle is ridden around a corner in the road the _____ the centripetal _____ acting on it.

2 Explain how an object can stay at a constant speed and at the same time change its velocity.

3 The photograph shows a space station in orbit around the Earth.

What happens to the centripetal force if:
i) the mass of the space station increases;
ii) the space station is moved into an orbit further from the Earth?

4   Under normal conditions the maximum speed that a lorry can go around a bend
    without skidding is 60 km/hr. Would the lorry be able to go around the bend
    faster, at the same speed, or would it need to slow down, if the road was muddy?
    Explain the reason for your answer.

Examination questions

1 a) Two sky-divers jump from a plane. Each holds
     a different position in the air.

   Copy and complete the following sentence.
   Sky-diver _____ will fall faster because
   _____   *(2 marks)*

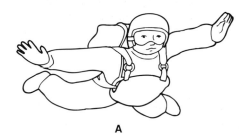

A

B

b) The diagram shows the direction of the forces
   acting on one of the sky-divers.

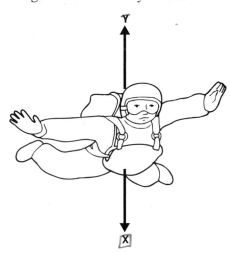

Copy the following sentences and complete them
by choosing the correct endings from the boxes.

i)   Force X is caused by

| air resistance |
| friction |
| gravity |

.

(1 mark)

ii) Force Y is caused by

| air resistance |
| gravity |
| weight |

.

(1 mark)

iii) When force X is bigger than force Y, the
     speed of the sky-diver

will

| go up |
| stay the same |
| go down |

.

(1 mark)

iv) After the parachute opens, force X

| goes up |
| stays the same |
| go down |

.

(1 mark)

c) How does the area of an opened parachute
   affect the size of force Y?   *(1 mark)*

2 Two students Anna and Graham took part in a
   sponsored run. The distance–time graph for
   Graham's run is shown. Four points have been
   labelled A, B, C and D.
   a) Between which pair of points was Graham
      running the slowest?   *(1 mark)*
   b) Anna did not start the run until 10 minutes
      after Graham. She completed the whole run at
      a constant speed of 4 m/s.

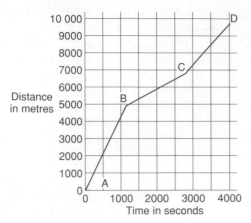

i) Write down the equation that links distance, speed and time. *(1 mark)*

ii) Calculate, in seconds, how long it took Anna to complete the run. Show clearly how you work out your answer. *(2 marks)*

iii) Copy the graph and draw a line to show Anna's run. *(2 marks)*

iv) How far had Graham run when he was overtaken by Anna? *(1 mark)*

3 Five forces, **A**, **B**, **C**, **D** and **E** act on the van.

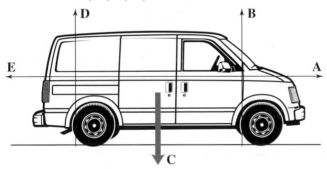

a) Copy and complete the following sentences by choosing the correct forces from **A** to **E**.
Force _____ is the forward force from the engine.
Force _____ is the force resisting the van's motion. *(1 mark)*

b) The size of forces **A** and **E** can change.
Copy and complete the table to show how big force **A** is compared to force **E** for each motion of the van.
Do this by placing a tick in the correct box.
The first one has been done for you.

Motion of van	Force A smaller than force E	Force A equal to force E	Force A bigger than force E
Not moving		✓	
Speeding up			
Constant speed			
Slowing down			

(3 marks)

c) When is force **E** zero? *(1 mark)*

d) The van has a fault and leaks one drop of oil every second.
The diagram below shows the oil drops left on the road as the van moves from **W** to **Z**.

W   **X**   **Y**   **Z**

Describe the motion of the van as it moves from **W** to **X**, **X** to **Y** and **Y** to **Z**. *(3 marks)*

e) The driver and passengers wear seatbelts. Seatbelts reduce the risk of injury if the van stops suddenly.

**backwards   downwards   force
   forwards   mass   weight**

Copy and complete the following sentences, using words from the list above, to explain why the risk of injury is reduced if the van stops suddenly.

A large _____ is needed to stop the van suddenly.
The driver and passengers would continue to move _____ .
The seatbelts supply a _____ force to keep the driver and passengers in their seats. *(3 marks)*

f) The van was travelling at 30 m/s. It slowed to a stop in 12 seconds. Calculate the van's acceleration. *(3 marks)*

4 The graph shows three stages of a van's journey.
a) During which stage of the journey **A–B**, **B–C** or **C–D**:
i) is the van stationary? *(1 mark)*
ii) is the van moving at a constant speed? *(1 mark)*

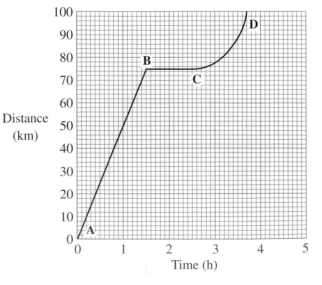

b) Calculate the gradient of the graph from
   A to **B**. *(2 marks)*
c) What does this gradient measure? *(1 mark)*

5 The diagram shows a parked car.

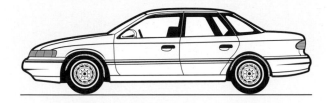

When the car is driven away, its engine gives a constant forward force.
The speed increases quickly at first, then more slowly. After a time the car reaches a constant speed.
Explain why the motion of the car changes in this way. *(3 marks)*

6 The diagram shows a spanner being used to undo a tight nut.

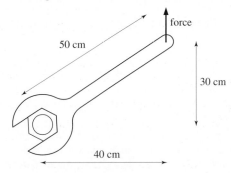

The nut was tightened using a moment of 120 newton metres.

Use the following equation to calculate the force needed to undo the nut. Show clearly how you work out your answer.

moment = force × perpendicular distance
from pivot

(2 marks)

7 The diagram shows a tower crane.

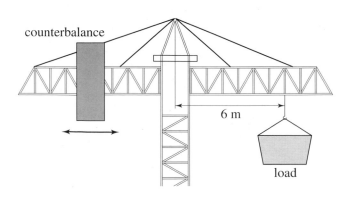

a) Explain why the crane would be unstable without the counterbalance. *(2 marks)*
b) The counterbalance can be moved to the left or right, as shown by the arrows on the diagram. Explain the advantage of having a movable counterbalance. *(2 marks)*
c) The load shown in the diagram is 75 000N. The load is 6 m from the tower. Calculate the turning effect (moment) of the load in newton metres. *(2 marks)*
d) The crane is balanced and horizontal. What is the turning effect (moment) of the counterbalance in newton metres? Explain your answer.

(3 marks)

8 a) A thin sheet of cardboard is cut to the shape below. Describe, with a diagram, an experiment to find its centre of mass.

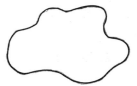

(5 marks)

b) Copy and label with an **X** the centre of mass of each of the three objects below.

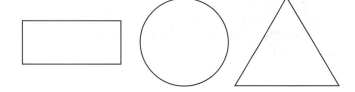

(3 marks)

c) Explain why a mechanic would choose a long spanner to undo a tight nut. *(2 marks)*

9 a) The diagram shows three aeroplanes at an airport.

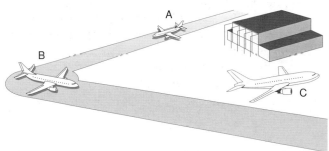

Aeroplane **A** is moving at constant velocity towards the main runway.
Aeroplane **B** is stationary, waiting to take off.
Aeroplane **C** has just taken off and is accelerating.

i) Which, if any, of the aeroplanes has zero momentum? *(1 mark)*

ii) The momentum of **one** of the aeroplanes is changing. Which one? Give a reason for your answer. *(2 marks)*

10 The picture shows luggage which has been loaded onto a converyor belt.

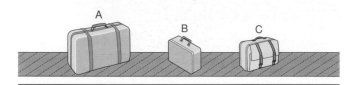

Each piece of luggage has a different mass.

mass of **A** = 22 kg
mass of **B** = 12 kg
mass of **C** = 15 kg

a) i) What is the momentum of the luggage before the conveyor belt starts to move? Give a reason for your answer. *(2 marks)*

ii) When the conveyor belt is switched on the luggage moves with a constant speed. Which piece of luggage **A**, **B** or **C** has the most momentum? Give a reason for your answer. *(2 marks)*

iii) At one point the conveyor belt turns left. The luggage on the belt continues to move at a constant speed.

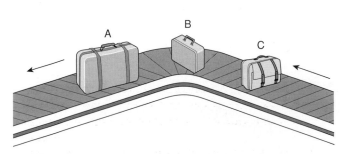

Does the momentum of the luggage change as it turns left with the conveyor belt? Give a reason for your answer.
(2 marks)

b) Which of the following units can be used to measure momentum?

J/s   kg m/s   Nm   *(1 mark)*

11 a) The diagram shows a simple design for a space rocket.

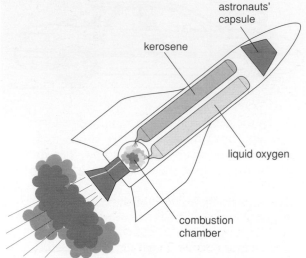

i) Explain, using the idea of momentum, how the initial propulsion of a rocket is produced. *(3 marks)*

ii) State and explain **one** way the acceleration of a rocket can be increased. *(2 marks)*

iii) In what unit is momentum measured? *(1 mark)*

b) The diagram shows an astronaut working in space. Releasing compressed gas from the back pack allows the astronaut to move around.

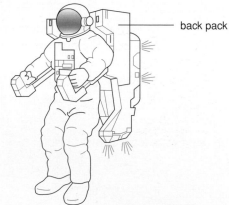

During one spacewalk, 0.5 kilograms of gas was released in 2 seconds. The gas had a speed of 60 metres per second. Use the following equation to calculate the force, in newtons, exerted on the astronaut by the gas. (Ignore the change in mass of the back pack).

$$\text{force} = \frac{\text{change in momentum}}{\text{time}}$$

(2 marks)

12 a) The picture shows two ice hockey players skating towards the puck. The players, travelling in opposite directions, collide, fall over and stop.

player 3
mass = 75 kg
speed = 4 m/s

player 4

i) Use the following equation and the data given in the box to calculate the momentum of player number **3** before the collision. Show clearly how you work out your answer and give the unit.

momentum = mass × velocity

(3 marks)

ii) What is the momentum of player **4** just before the collision? *(1 mark)*

iii) The collision between the two players is **not** *elastic*. What is meant by an *elastic* collision? *(1 mark)*

b) The pictures show what happened when someone tried to jump from a stationary rowing boat to a jetty. Use the idea of momentum to explain why this happened.

(2 marks)

c) The diagram shows one type of padded body protector which may be worn by a horse rider.

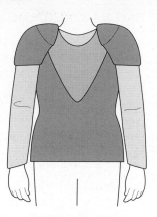

If the rider falls off the horse, the body protector reduces the chance of the rider being injured. Use the idea of momentum to explain why. *(3 marks)*

13 The diagram shows a satellite in orbit around the Earth.

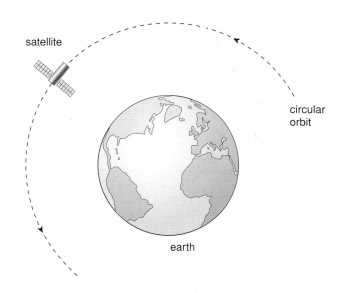

satellite

circular orbit

earth

a) Copy and complete the diagram, drawing an arrow on the diagram to show the direction of the centripetal force which acts on the satellite. *(1 mark)*

b) Use words from the following list to complete the sentences.

greater   less   unchanged

i) If the mass of the satellite decreases then the centripetal force needed is ———— .

ii) If the speed of the satellite increases then the centripetal force needed is ———— .

iii) If the radius of the orbit increases then the centripetal force needed is ———— .

(3 marks)

14 The following paragraphs appeared in a newspaper.

> ## JEEP FAILS GOVERNMENT TEST
>
> A car manufacturer confirmed yesterday that one of its four-wheel drive mini-jeeps rolled over at 38 mph during stability tests conducted by the Government.
>
> Testing has been halted until safety cages can be fitted to seven makes of mini-jeeps which the Department of Transport has agreed to test after repeated complaints from the Consumer's Association. The Association claims its own tests show that the narrow track, short-wheelbase vehicles are prone to rolling over. The Government's tests highlight how passengers raise the centre of mass. All seven vehicles passed the test unladen, although one raised two wheels.

a) Write down **two** factors mentioned in the newspaper article which affect the stability of vehicles. *(2 marks)*

b) The diagram shows a tilted vehicle.

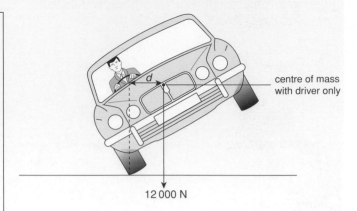

centre of mass with driver only

12 000 N

The distance **d** shown in the diagram is 50 cm. Calculate the moment of the force about the point of contact with the road. *(3 marks)*

c) Explain how passengers make the vehicles more likely to roll over (less stable). You may use diagrams if you wish. *(4 marks)*

Chapter 3
Waves

3.1		
Co-ordinated	**Modular**	
10.13	Mod 12 13.1	

Characteristics of waves

If a rope is shaken up and down a wave travels along it. A person holding the other end of the rope will feel the pulse when it arrives. The pulse has carried energy but when the pulse has passed, the rope remains exactly as it was before. None of the material of the rope has moved permanently. So **waves** carry energy from one place to another without transferring matter.

Figure 3.1
A wave moving along a rope

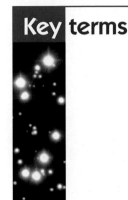

movement of rope

movement of wave

As the pulse passes, the rope moves up and down. The movement of the rope is at right angles to the movement of the pulse. This is an example of a **transverse wave**.

Energy can also be transmitted as a pulse in a spring by shaking it backwards and forwards.

When the spring is shaken backwards and forwards, each coil moves backwards and forwards rather than up and down. The movements are in the same direction that the energy travels. In some places the coils bunch together (areas of **compression**) and in other places they are further apart (areas of **rarefaction**). This type of wave is called a **longitudinal wave**.

Figure 3.2
A wave moving along a spring

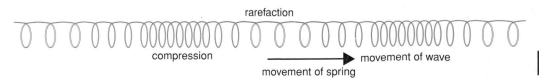

rarefaction

compression

movement of wave

movement of spring

Waves

Light and sound travel as waves. Energy is transmitted but there is no movement of mass. Sound waves travel through solids, liquids and gases as longitudinal waves. Most waves, for example water waves, waves in a rope and light waves, are transverse waves. Light waves can travel through a vacuum.

For any wave the number of waves made each second is called the **frequency**. Frequency is measured in units called **hertz** (Hz). One hertz is one **vibration** or **cycle** per second. So 2000 vibrations in a second would be described as 2000 Hz or 2 kHz (kilohertz).

For any wave system there are two factors needed to describe the wave. For a transverse wave these are the distance between the **crests** or the distance between the **troughs** of a wave – which is called the **wavelength** – and the height of the wave crest, which is called the **amplitude**.

Figure 3.3
A transverse wave in water showing the wavelength and amplitude

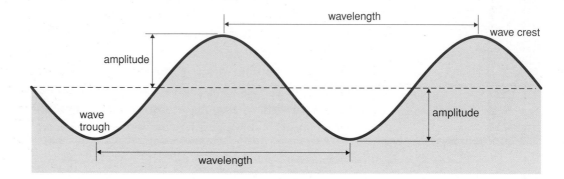

The amplitude depends upon the energy in the wave. The greater the amplitude the greater the energy of the wave.

The behaviour of waves

Waves and refraction

Light can pass through transparent materials like air, water, glass and perspex. When light crosses the boundary between two transparent materials at an angle greater than 0° to the normal it changes direction – this is called **refraction**.

When light goes from one material to another it changes speed. This change of speed causes the light to change direction. Light travels slower in glass than it does in air.

The diagram shows that when light travels from air to glass it bends towards the **normal** on its way into the glass because it slows down. When light travels from glass to air it bends away from the normal on its way out of the glass because it speeds up.

Figure 3.4
Refraction of light through a glass block

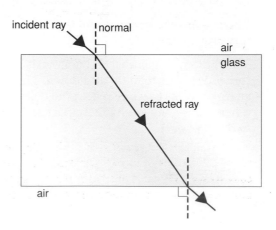

When light changes speed as it moves from one transparent substance to another, the frequency does not change but the wavelength does.

Water waves also show refraction effects. Figure 3.5 shows water waves being slowed down as they pass over an area of shallow water. They are not being refracted because all parts of the wave meet the edge of the shallow water at the same time. When the waves leave the shallow water they speed up again.

Figure 3.6 shows the water waves hitting the shallow water at an angle. This time the waves are not only slowed down but they have changed direction. The water waves have been refracted.

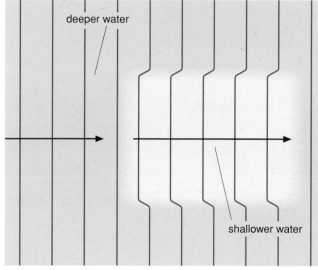

Figure 3.5

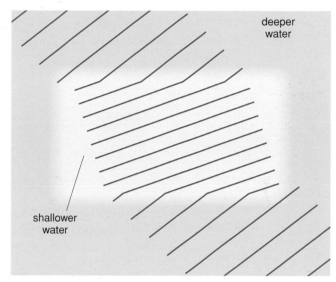

Figure 3.6

Waves and diffraction

Figure 3.7 shows water waves passing through gaps of different sizes. When they pass through the gap that is about the same width as their wavelength they spread out as they emerge on the other side. If the gap is very wide compared to the wavelength then very little spreading out occurs. Waves will also spread out when they pass the edge of an object. This spreading out of a wave as it passes the edge of an object or as it moves through a gap is called **diffraction**.

Figure 3.7
Diffraction of waves at a narrow and at a wide gap

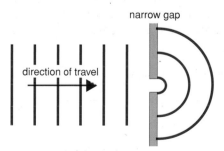

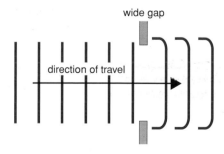

Therefore the degree of diffraction depends on the wavelength of the wave and the width of the gap it passes through. Because of this, some diffraction effects are not easy to see. Visible light with its very short wavelengths (0.0000005 m) would need a gap with a width to match the wavelength in order for diffraction effects to be seen.

Waves

VHF radio waves have wavelengths of a few metres, whereas long wave radio waves have wavelengths of thousands of metres. Long wave radio signals are readily diffracted by the large gaps between buildings and as they pass over hills. This means that long wave radio signals can be received in the shadow of a hill. VHF waves are less likely to show diffraction effects and spread out, so the receiving of these signals is difficult in hilly areas.

Figure 3.8
The diffraction of
a) VHF waves
b) long wave radio
waves

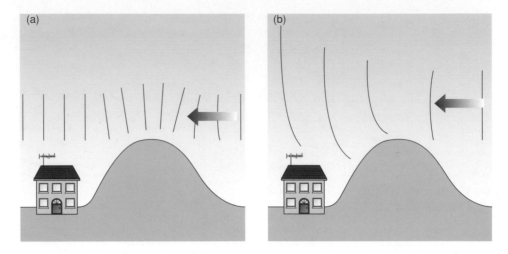

Sound waves have wavelengths of about one metre so they can readily spread out as they pass through an open doorway or as they hit an obstacle such as the corner of a building. So someone outside a room may hear a conversation although they cannot see the speakers.

Figure 3.9
Diffraction of sound
waves through a
doorway

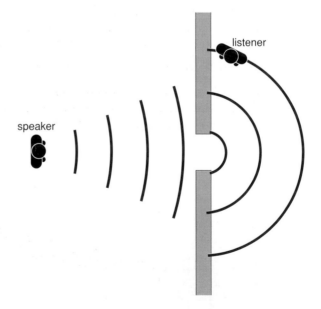

Total internal reflection

When a ray of light emerges from glass, perspex or water into air it bends away from the normal. As the internal angle gets bigger so does the external angle. When the internal angle reaches about 42°, the external angle reaches 90°. When this happens the internal angle is called the **critical angle**. So for glass the critical angle is 42°.

If a ray of light hits the inside of a glass block at an angle above the critical angle it reflects inside the block. Because ALL the light reflects inside the block, this is called **total internal reflection**.

Figure 3.10
Demonstrating total internal reflection

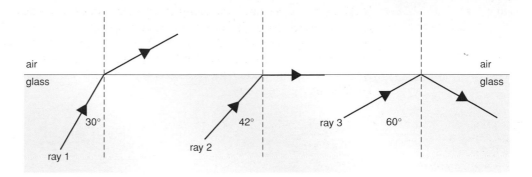

Light entering the prism hits the first angled face at 45°. This angle is greater than the critical angle so the light ray is totally internally reflected. The ray then hits the second face at 45°. Once again the ray is totally internally reflected. It comes out of the prism parallel to the direction it went in.

Figure 3.11
Total internal reflection inside a 45°/45°/90° glass prism

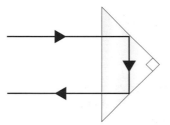

Prisms are used in 'prismatic' binoculars because the quality of the image is much better with a prism than a mirror. The reflectors on cars and bicycles and in some 'cat's eyes' in the road use total internal reflection in a 45°/45°/90° prism.

A more recent application of total internal reflection is in fibre optics.

Fibre optics allow light rays entering one end of the fibre to be totally internally reflected numerous times until they leave at the other end. No matter how much the fibre optic gets twisted or bent. the light will always pass all the way along the fibre.

Figure 3.12
Passage of light through fibre optic

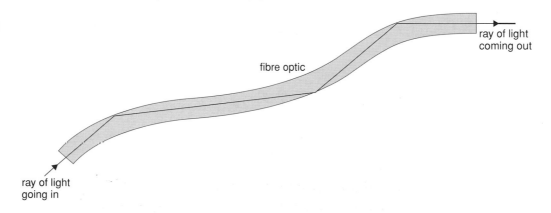

Total internal reflection in fibre optics has two very important practical uses:

- *Medical uses.* An endoscope is a bundle of thin optical fibres which can be inserted into the body. Light is sent down some fibres and reflected back so that the medical observer can view the inside of the patient. This has enabled 'key-hole' surgery to be developed where very small incisions can be made rather than using more invasive surgery.

Figure 3.13
An endoscope in use

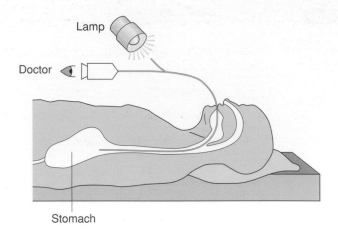

Lamp

Doctor

Stomach

- *Communications* e.g. cable telephones and cable television. The use of fibre optic cable means that as many as 500 000 telephone conversations can be carried at the same time. This is hundreds of times more than can be carried by a copper cable of similar size. The fibre optic cable is also cheaper.

Summary

- Waves transfer energy not matter.

- The **amplitude** of a transverse wave is the maximum distance of the wave above or below the middle position.

- The **wavelength** of a transverse wave is the distance between the tops of successive crests (or the bottoms of successive troughs).

- The **frequency** is the number of waves produced in 1 second. It is measured in **hertz** (Hz).

- In a **transverse** wave the disturbance causing the wave is at right angles to the direction in which the energy travels.

- In a **longitudinal** wave the disturbance causing the wave is in the same direction in which the energy travels.

- **Refraction** is caused by changes in the speed of a wave as it passes through materials with different densities.

- When waves move through gaps or pass the edges of objects they can spread out. This effect is called **diffraction**.

- Under certain conditions light can be totally internally reflected.

Topic questions

1 Draw a water wave and mark on it the wavelength and the amplitude.

2 a) Which of the following is the unit of frequency?
   amp   hertz   metre
   b) What is meant by the frequency of a wave?

3 a) What are longitudinal waves?
   b) Give an example of a longitudinal wave.

4 a) What are transverse waves?
   b) Give two examples of transverse waves.

5 a) What is the value of the critical angle for a ray of light in a glass block?
   b) When will a ray of light show total internal reflection?

6 a) What is diffraction?
   b) What happens to the speed of a water wave as it passes over shallow water?
   c) Why is it difficult to show that light waves can be diffracted?
   d) Why can sounds easily spread around corners?

3.2 The wave equation

Co-ordinated	Modular
10.13	Mod 12 13.1

Imagine a wave with a wavelength of λ metres travelling with a **wave speed** of v metres per second.

In 1 second the wave travels v metres.

Each wave has a length of λ metres so in v metres there are $\frac{v}{\lambda}$ waves.

So in 1 second there are $\frac{v}{\lambda}$ waves.

This means the frequency of the wave (f) is $\frac{v}{\lambda}$ Hz so $f = \frac{v}{\lambda}$

This equation can be rewritten as:

wave speed   = frequency × wavelength
(metre/second, m/s)   (hertz, Hz)   (metre, m)

$$v = f \times \lambda$$

Example: What is the speed of waves which have a frequency of 200 Hz and a wavelength of 0.5 m?

frequency = 200 Hz
wavelength = 0.5 m
So, wave speed = 200 × 0.5 m/s = 100 m/s

Summary

◆ **Wave speed**   =   frequency × wavelength
   (m/s)   (Hz)   (m)

Topic questions

1 Draw two waves, one of which has double the wavelength and half the amplitude of the other.

2 What is the speed of a sound wave having a frequency of 512 Hz and wavelength 0.6 m?

3 Radio 4 has a wave speed of 300 000 000 m/s and a wavelength of 1500 m. What is its frequency?

3.3 The electromagnetic spectrum

Co-ordinated	Modular
10.14	Mod 12 13.2

The **electromagnetic spectrum** contains a range of transverse waves all of which can travel through a vacuum at the same speed (300 000 000 m/s, or 3×10^8 m/s). These waves are classified into several groups according to their wavelength and frequency and how they behave. Light is one of these groups.

Because each part of the electromagnetic spectrum has a different wavelength and frequency, each part will be reflected, refracted, absorbed or transmitted differently.

Figure 3.14
Wavelengths and frequencies of the electromagnetic spectrum

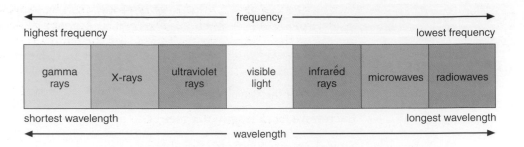

When electromagnetic radiation is absorbed, the energy it carries is likely to make the absorbing object hotter, it may even produce an alternating current with the same frequency as the radiation.

Uses and dangers of different parts of the electromagnetic spectrum

Radio waves

Radio waves have wavelengths between about 1500 metres (long-wave radio signals) and a few metres (very high frequency radio signals – VHF signals). Television transmissions use even shorter wavelengths (about 0.1 metres). Radio waves are used for communications.

The longer wavelength radio waves are reflected from an electrically charged layer high up in the Earth's atmosphere. This enables them to be sent between distant points despite the curvature of the Earth's surface.

Microwaves

Microwaves are also used for communication purposes. They have shorter wavelengths than radio waves and therefore have a greater frequency. This enables them to carry more information. Microwaves are used for mobile phones and for carrying television channels between city centres via very tall telecommunication towers.

Microwaves can penetrate the upper atmosphere so can be sent to satellites and back to Earth. Many people receive some of their television programmes via microwaves which have travelled from the Earth to a satellite and then back to a dish aerial on the side of their house.

Figure 3.15
Sending a signal via a satellite

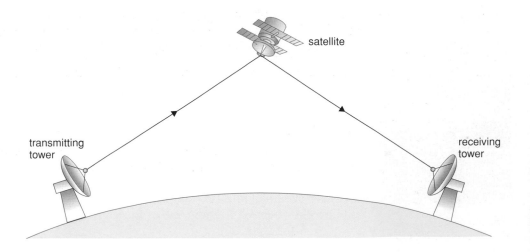

Microwaves can also be used to cook food. The microwave's energy is absorbed easily by water molecules in the food. This makes the water molecules vibrate more, so increasing their kinetic energy. In this way the temperature is raised and the food cooks very quickly. In a microwave oven liquids heat up more quickly than solids. This is why a microwaved doughnut may only be warm on the outside but have scalding hot jam inside. Microwaves can seriously damage living tissue. Microwave ovens have to be very carefully screened to prevent radiation escaping and the oven will turn off immediately if the door is opened.

Infrared radiation

Infrared radiation is used in the remote controls for televisions and stereo systems.

It is the heat radiated from grills, toasters and radiant heaters.

Infrared waves are absorbed by skin giving a sense of warmth but excessive exposure will lead to skin burns.

> **Did you know?**
>
> Infrared cameras can detect objects at different temperatures and are used by the army and police to spot people in the dark. Rescue teams use infrared cameras to detect living people in rubble after explosions.

Ultraviolet radiation

Ultraviolet rays are responsible for sun tans. The ozone in the upper atmosphere shields us from an excess of ultraviolet, but there are concerns that 'holes' are developing in the upper ozone layer due to the effects of CFCs.

Ultraviolet radiation is dangerous because the energy can penetrate living tissue and may cause skin cancer. So it is important that appropriate creams are used to protect the skin from this radiation.

Ultraviolet rays create visible light in fluorescent lamps. In these lamps the ultraviolet rays transfer their energy into light by hitting fluorescent coatings on the inside of the glass.

Ultraviolet radiation is also used in security coding. Numbers or names are written on to valuable objects using special ink that is visible only in ultraviolet light.

X-rays

X-rays affect a photographic plate causing it to be exposed. Medical uses of X-rays rely on the fact that living tissue is almost transparent to X-rays but bones absorb most of the rays. It is thus possible to study broken bones with X-rays.

X-rays are also useful for security scanning at airports because metal objects stop the X-rays.

Gamma rays

Gamma rays (see section 6.1) are used to kill harmful micro-organisms in food, sterilise surgical instruments and kill cancer cells.

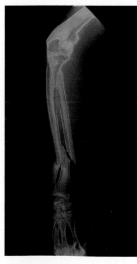

Figure 3.16
An X-ray of a broken arm

Dangers of exposure to electromagnetic radiation

It is important to realise that although each of the waves of the electromagnetic spectrum can be used for beneficial purposes there are many dangers that could arise from their misuse. Many of these dangers affect the cells of living organisms, including humans. It is necessary therefore that measures be taken to reduce the chances of exposure to many types of radiation. Figure 3.17 outlines some of the dangers caused by over exposure to the different types of radiation and some of the measures taken to reduce such exposure.

Figure 3.17
The effects of different types of radiation on living cells

Radiation	Effect
microwave	These are absorbed by water in the cells, so cells can be damaged or killed by the heat produced.
infrared	These are absorbed by the skin and felt as heat which can damage or kill cells.
ultraviolet	These can pass through the skin and reach the deeper tissues. This radiation causes the skin to darken and can cause skin cancers. The darker the skin the more ultraviolet radiation it absorbs and less reaches the deeper tissues. Over-exposure to ultraviolet radiation should be avoided and the use of various sun-creams that have been designed to protect the skin from this radiation is recommended.
X-rays and gamma radiation	Although most of these pass through soft tissue some of their energy is absorbed causing cells to become cancerous. Workers who are likely to be operating machinery that produce X-rays and gamma radiation wear protective clothing that includes lead shielding to protect their reproductive organs from excessive exposure. Workers who are likely to come in contact with gamma radiation and other forms of radiation from radioactive sources wear special badges that contain photographic film. This film is regularly developed to check whether any over exposure to radiation has occurred.

Figure 3.18▼
Using X-rays to treat a cancerous tumour

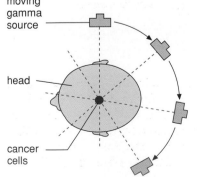

Radiation and cancers

Although low level doses of ultraviolet radiation, X-rays and gamma radiation can cause normal cells to become cancerous, high doses of these types of radiation can be used to kill cancerous cells.

The cancer is exposed to X-rays projected from several directions. This makes sure that the beams are concentrated in the area of the cancer and that the surrounding cells receive only a low dose of the radiation.

With gamma ray treatment the source moves around the cancer. Again, this makes sure that the radiation is concentrated in the area of the cancer and that the surrounding cells receive only a low dose.

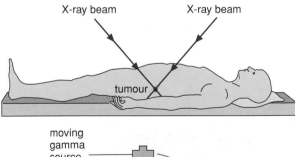

Figure 3.19▲
Using gamma rays to treat a cancerous tumour

Sending messages

The telegraph was the first invention to use an electric current as a message carrier. Morse code was invented for this first message carrier. Morse code is made up from short and long pulses. These pulses were the first form of **digital signals**. Digital signals have only two states, 'on' or 'off'.

Figure 3.20
Morse code as a series of digital signals

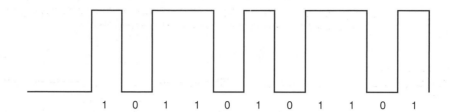

The telephone, invented in 1876, consists of a microphone and a loudspeaker. The microphone changes sound waves into electrical signals that can be displayed as transverse waves on a CRO (see section 3.5). These signals are called **analogue signals**. Analogue signals are continuous waves that vary in amplitude and frequency.

Figure 3.21
Analogue signals produced by a microphone

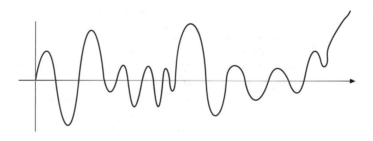

The radio was invented in 1901 and for many years radio signals were transmitted as analogue signals. One form of these signals consisted of a carrier wave sent at a fixed frequency. The signal being carried was sent by changing the carrier wave's amplitude to match the amplitude of the message being transmitted.

Figure 3.22
Carrier wave and signal wave

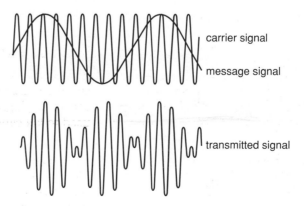

Today the air is crowded with radio signals and the telephone wires can hardly cope with the number of conversations going on. Optical fibres (see section 3.1) are being used to replace the main telephone cables. These optical fibres will carry coded pulses of light instead of electrical signals. A standard telephone copper cable can carry up to about 10 000 coded conversations at any one time. An optical fibre can carry at least 110 times as many conversations coded as pulses of light.

The digital radio signals or the digital pulses of light do have certain advantages over analogue signals.

- The information carried as a digital signal is unlikely to get changed – the signal is either 'on' or 'off'. This means the signals received are of a high quality and sound like the original.

- More information in a given time can be transmitted because digital signals take up less wave space than analogue signals.

As analogue signals travel from the transmitter circuit to the receiver circuit they gradually weaken and often collect unwanted signals called 'noise'. Different frequencies of an analogue signal weaken by different amounts, so during any amplification process to make the weak signal stronger, the original signal gets more and more distorted.

Digital signals will also weaken as they travel from the transmitter to the receiver. However they are still recognisable as 'on' or 'off' states. Any unwanted 'noise' usually has a very low amplitude so is interpreted by the receiver circuit as 'off'.

Summary

◆ The **electromagnetic spectrum** is a range of transverse waves that travel at 300 000 000 m/s in a vacuum.

◆ Electrical messages can be sent either as **digital signals** or as **analogue signals**.

Topic questions

1 Use the types of electromagnetic radiation in the box to answer the following questions. Some of the questions have more than one correct answer.

| radio   microwaves   infrared   light   ultraviolet   X-rays   gamma rays |

Which type(s) of radiation can be used a) for communications b) for cooking c) to give a sun tan d) cause cancer e) to sterilise equipment?

2 Why are microwaves so dangerous to living tissue?

3 a) What are analogue signals? b) What are digital signals?

4 Why are copper telephone cables being replaced by optic fibre cables?

3.4   Optical devices

Co-ordinated	Modular
10.15	Mod 23 14.7

Lenses

A lens refracts (bends) light to produce an image. Many optical devices such as cameras, magnifying glasses, spectacles (glasses) and microscopes use lenses. There is even a lens in each of your eyes.

Although lenses can be made from any transparent material, they are usually made from glass. The glass lenses can be ground and polished to remove any small imperfections on their surfaces. Most, but not all, lenses have two spherical or nearly spherical surfaces.

There are two main types of lens, converging (or convex) and diverging (or concave).

Converging lenses

In general any lens that is thicker in the middle than it is at the edges is a **converging lens**.

Figure 3.23
Converging lenses

Light entering the lens is refracted towards the normal. Light leaving the lens is refracted away from the normal.

Figure 3.24
Light refracted by a converging lens

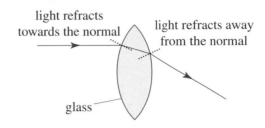

As a result of refraction, parallel rays of light entering the lens will converge to meet at a single point. This point is called the **focus** of the lens. The thicker the lens the closer the focus is to the lens.

Figure 3.25
Parallel rays of light passing through a convex lens converge at the focus

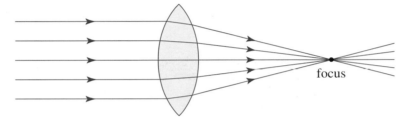

Light can pass through the lens in either direction. So parallel rays of light coming from the right would converge to a point at an equal distance to the left of the lens.

A converging lens can focus the light from a distant object onto a sheet of paper. An image of the object will be seen on the paper. The image will be sharpest when the paper is positioned at the focus of the lens.

Figure 3.26
A converging lens can be used to project an image onto a piece of paper

This sort of image is called a **real image**. When the light from an object passes through its image then it is a real image. A real image can be shown on a screen. In a camera, a converging lens is used to produce a real image of an object on photographic film. The image is smaller than the object and nearer to the lens than the object.

Figure 3.27
A camera can be used to produce a real image on photographic film

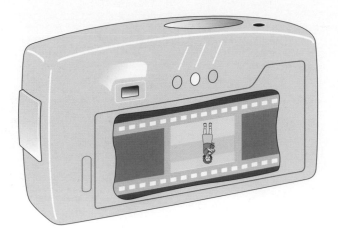

Figure 3.28
The image produced is both smaller and nearer to the lens than the object is

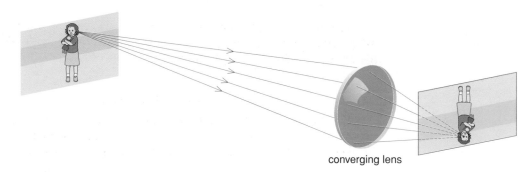

converging lens

Diverging lenses

In general any lens that is thicker at the edges than it is in the middle is a **diverging lens**.

Figure 3.29
Diverging lenses

Light entering the lens is refracted towards the normal. Light leaving the lens is refracted away from the normal.

Figure 3.30
Light refracted by a diverging lens

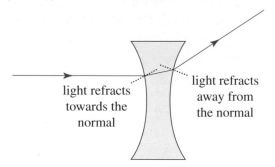

light refracts towards the normal

light refracts away from the normal

As a result of refraction, parallel rays of light entering the lens will spread out or diverge as they leave the lens. The rays do not meet at a single point but they do look as though they diverge from a single point. This point is the focus of the lens. It is a virtual focus.

Figure 3.31
The virtual focus of a diverging lens

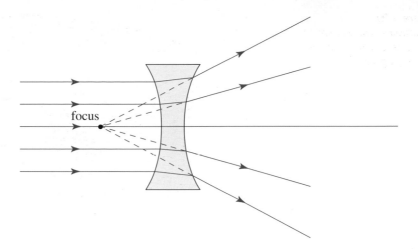

A diverging lens always produces a **virtual image**. It is impossible to show a virtual image on a screen as the light does not actually pass through it. The only way to see the image is to look through the lens.

Ray diagrams for converging lenses

The position and nature of an image can be found by drawing a ray diagram. Two rays of light are drawn from one point on the object. Each ray of light always takes a fixed path. A real image is formed where the two rays of light cross.

It is usual to show the refraction as if it all happens at a straight line that runs through the centre of the lens. This is a simplification as refraction really happens at both surfaces of the lens.

A ray of light parallel to the axis of the lens is refracted through the focus.

Figure 3.32

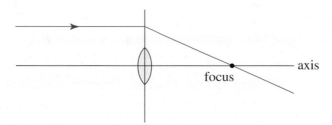

A ray of light going through the centre of the lens continues in a straight line. (Strictly this is not true, but it is a useful approximation).

Figure 3.33

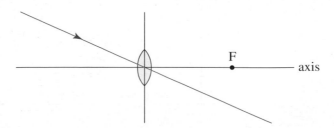

The position of the object affects the image.

117

Figure 3.34
The image produced for objects far from the lens

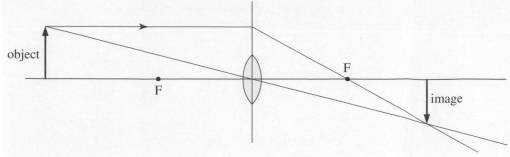

The image is:

● real,
● inverted (upside down),
● smaller than the object.

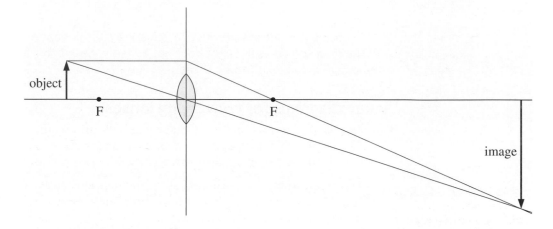

Figure 3.35
The image produced for objects close to the lens

The image is:

● real,
● inverted,
● bigger than the object (magnified).

Ray diagrams for diverging lenses

These are drawn using the same two rays of light as for a converging lens. The ray of light going through the centre of the lens will continue in a straight line. But the ray of light parallel to the axis is refracted so that it seems to have come from the virtual focus. A virtual image is formed where the two rays seem to cross.

Figure 3.36
The image produced by a diverging lens

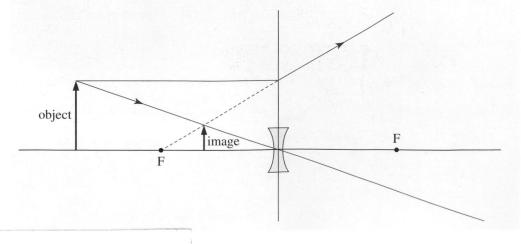

- virtual,
- upright,
- smaller than the object.

The image produced by a diverging lens is always virtual, upright and smaller than the object.

The magnifying glass

A single converging lens can be used as a magnifying glass. The most powerful magnifying glasses use very thick lenses.

Figure 3.37
A converging lens produces a magnified image of a close object

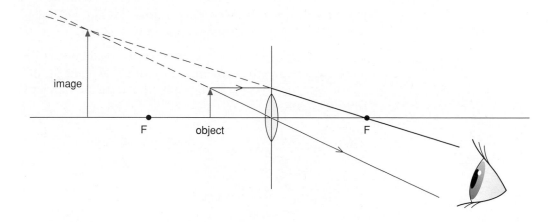

Figure 3.38
When a person looks through a converging lens at a close object they see a magnified image of that object

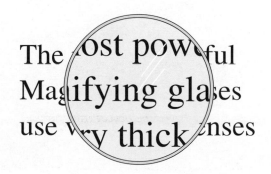

When the object is between the lens and its focus, the rays of light from the object passing through the lens do not cross over. The only way of seeing the image is to look through the lens. The image is a virtual image. It is also upright and magnified.

> **Did you know?**
> The first person to see microbes in water was Anton van Leeuwenhoek (1632–1723). He made a perfectly smooth, spherical lens from a grain of sand.

The camera

A camera uses a converging lens to produce an inverted, small, real image on photographic film.

Figure 3.39
A cross section of a camera

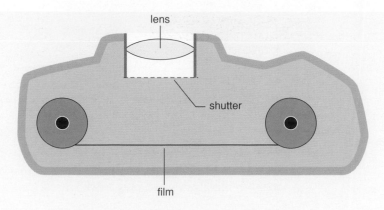

Light reaches the film when the shutter is open. The image of a fast moving object will be blurred if the shutter is open for too long.

To focus a camera the lens is moved towards or away from the film. This movement allows the camera to produce sharp images of close and distant objects. The closer the object is to the camera, the further the lens needs to be moved away from the film.

Figure 3.40
A camera lens

Summary

◆ A converging lens can produce a real or a virtual image.

◆ A diverging lens always produces a virtual image.

◆ A real image can be shown on a screen, a virtual image cannot.

◆ A camera uses a converging lens to produce a real image on photographic film.

◆ A converging lens can produce an upright, magnified, virtual image.

Topic questions

1 Copy and complete the following sentences:
  a) A _____ lens is thinner in the middle than at the _____.
  b) Parallel rays of light _____ by a convex lens will meet at the
     _____ .
  c) A camera produces an _____ , _____ image of
     an object.

2 Which of the lenses below are converging?

3 What type of image does a diverging lens always produce?

4 Describe a method to find the approximate distance between a converging lens
   and its focus.

5 Copy and complete each of these ray diagrams. 'F' marks the focus of the lens.

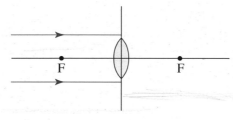
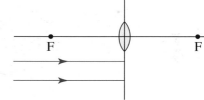

(a)   (b)

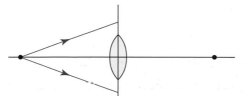
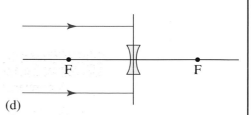

(c)   (d)

6 In a camera a lens is used to produce an image of an object on the film.
  a) What type of lens is used to produce the image in a camera?
  b) Give two ways in which the image produced on the film is different from the
     object.

7   Copy and complete the following ray diagram to find the position, size and type of image.

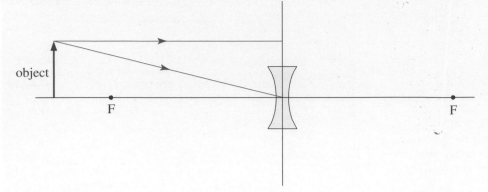

object

F

F

8   A converging lens is used as a magnifying glass. The distance between the centre of the lens and the focus of the lens is 5cm. An object 2cm high is placed 3cm from the lens.
a) Draw a ray diagram to find the position, size and type of image.
b) What magnification has the lens produced?

3.5   Sound and ultrasound

Co-ordinated	Modular
10.16	Mod 12 13.5

Sound is not an electromagnetic wave. Sound waves differ from electromagnetic waves in many ways.

Sound is produced whenever an object vibrates at a frequency which the ear can detect. The vibrating object creates very small pressure waves in the air. These pressure waves are longitudinal waves so the air is alternately compressed and rarefied. Each air particle vibrates but does not move permanently along the sound wave. The energy of a sound wave is passed from one particle to another by collision.

Sound waves cannot pass through a vacuum as there are no particles to carry the vibrations.

rarefaction

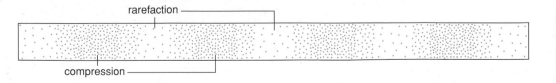

Figure 3.41
Variation in air pressure in a sound wave

compression

Sound can travel through solids and liquids as well as through gases like air. The greater the density of the medium, the faster the sound travels. Thus sound travels faster through water than it does through air. This is because in water the particles are closer together so can pass on their vibrations quicker. Sound travels even faster through solids such as wood because the particles are even closer together.

Looking at sound waves

A microphone transfers sound energy into electrical energy. If the microphone is connected to a cathode ray oscilloscope (CRO) a wave is displayed on the screen.

The shape of the wave shows how the air pressure at the microphone changes with time. It represents the sound wave but it is not a picture of a sound wave. Remember, sound waves are longitudinal **not** transverse. Like all waves, sound waves have amplitude and frequency.

Figure 3.42
Displaying sound waves on an oscilloscope screen

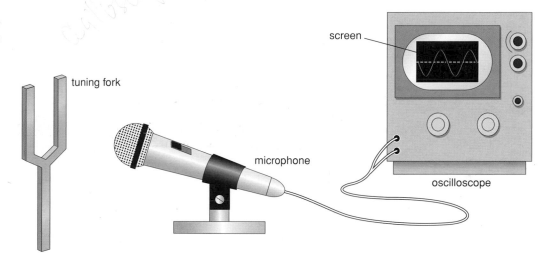

The amplitude of a sound wave is a measure of its loudness. The louder the sound, the bigger the difference in pressure between the areas of compression and the areas of rarefaction in the sound wave. This means that loud sounds carry more energy. On the CRO screen, the louder the sound the higher the crests and deeper the troughs of the wave on the screen.

The frequency of a sound wave is a measure of its pitch. The higher the frequency, the higher the pitch. A higher frequency means the wavelength is shorter. So on a CRO screen, a higher pitch sound means the waves are closer together and you will see more waves on the screen.

The traces shown in Figure 3.43 are for 'pure' or regular sounds. A tuning fork produces a reasonably pure sound.

Did you know?

If you double the frequency of a note, the note is exactly one octave higher in pitch.

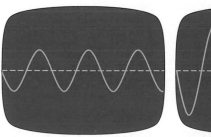

a) CRO trace of a sound wave

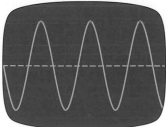

b) CRO trace of a sound with the same pitch as a) but much louder

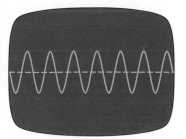

c) CRO trace of a sound with the same loudness as a) but a higher pitch

Figure 3.43
Waves on oscilloscope screens

Two different musical instruments playing the same note do not sound exactly alike. This is because the sound produced is not a pure sound. The diagram shows the trace of the sound of two different instruments playing the same note. The frequencies and amplitudes are the same but the shape of the note is different. This is what gives a musical instrument its particular sound (timbre).

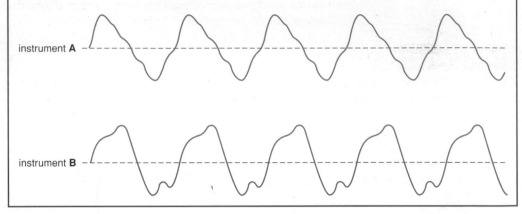

instrument **A**

instrument **B**

Ultrasound

Ultrasound is any sound which has a frequency or pitch which is too high for humans to hear, although other animals may be able to. Because of its short wavelength, the ultrasound produced from a small source is unlikely to spread out by diffraction.

Ultrasound can have a frequency as high as 150 000 Hz. At this high frequency the waves can travel through most substances as a narrow beam. Some of the ultrasound is always reflected back when the wave hits the boundary between two different types of materials. A detector placed near the source of the waves measures how long it takes for the waves to be reflected back by various boundaries. The time taken for the reflections to occur is used to generate a visual display.

Ultrasound can be used to produce the image of a **fetus**. The ultrasound passes directly through the skin of the mother into the fetus. The reflected sounds are converted into pictures and enable medical staff and prospective parents to get an early image of the developing baby. This is called **fetal imaging**. The advantage of ultrasound is that it does not harm living cells.

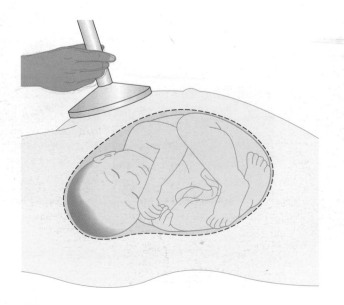

Figure 3.44
Using ultrasound in fetal imaging

Ultrasound can be used for cleaning delicate equipment without the need to take it apart. The ultrasound causes the dirt particles to vibrate rapidly and fall off. Ultrasound can also be used to check for cracks or defects in metal castings. The ultrasound waves are distorted by imperfections in the structure of the metal.

Summary

◆ **Ultrasound** (ultrasonic waves) has a frequency above the upper limit of the human hearing range (above about 20 000 Hz).

Topic questions

1 Ultrasonic echo sounders do not use a continuous signal. The ultrasound is sent out in a series of pulses with silence in between. Why is this necessary?

2 A CRO trace of a sound wave looks like this.
   a) The beam takes 0.005 s to travel across the screen. What is the frequency of the sound wave?

b) How would the trace differ if the sound
   i) had twice the frequency
   ii) was much quieter.

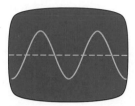

3.6		Seismic waves

Co-ordinated	Modular
10.17	Mod 12 / 13.6

The Earth is a layered structure with a **core**, a **mantle** surrounding the core and a thin solid **crust**.

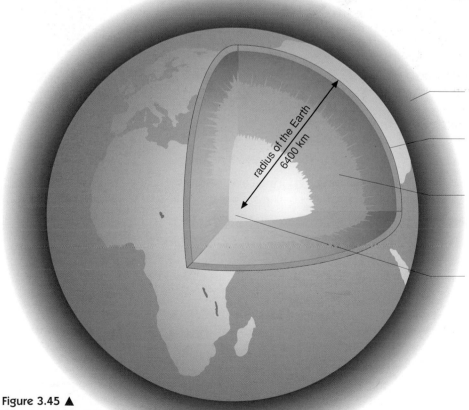

atmosphere:
a capsule of gases between -50 and 50°C

crust:
a thin layer of less dense solid rock between -50 and 1500°C

mantle:
a thick layer of moderately dense solid and molten rock between 1500 and 4000°C. It has all the properties of a solid except that because it is hot it can flow slowly

core:
a central ball of very dense, hot molten nickel and iron at 4000°C. The outer part of the core is liquid, the inner part is solid

The **lithosphere** is made up of the crust and the solid upper layer of the mantle

radius of the Earth 6400 km

Figure 3.45 ▲
The structure of the Earth

Waves

Evidence for this layered structure comes from studying the density of the Earth and analysing the movement of **seismic waves**.

The density of the Earth

Because the overall density of the Earth is much greater than that of the materials making up the crust, the interior of the Earth must be made up of different materials to the crust. These materials must be much denser than the materials in the crust. The high density of the interior can be explained if the **core** is considered to be made mainly of metals.

Seismic waves and the Earth

Earthquakes are caused by shock waves created when two parts of the Earth's crust move relative to each other. The shock waves are called seismic waves. The waves are detected using **seismographs**.

Figure 3.46
Quakes recorded on a seismograph

Two types of seismic waves are P waves and S waves.

- **P (primary) waves.** These are longitudinal and can pass through solids and liquids. They are called primary waves because they travel faster and are detected first.

- **S (secondary) waves.** These are transverse waves which can only travel through solids. S waves travel more slowly than P waves.

It is easy to remember which waves are longitudinal and which are transverse. P waves are Pushing and Pulling waves (longitudinal) and S waves are Shaking Sideways waves (transverse).

Although both waves travel at different speeds their speed increases with depth. Because the density of each layer of the Earth changes gradually the speed of each wave changes gradually and so the waves follow curved paths. As the waves pass from one layer to another the waves are refracted (bent).

The outer core is liquid so only P waves can pass through it. (Even though the inner core is solid, only P waves pass through it because it is completely surrounded by liquid.) The mantle is solid and allows both waves to pass through it.

Figure 3.47
A simplified diagram to show the way that P waves and S waves travel through the Earth

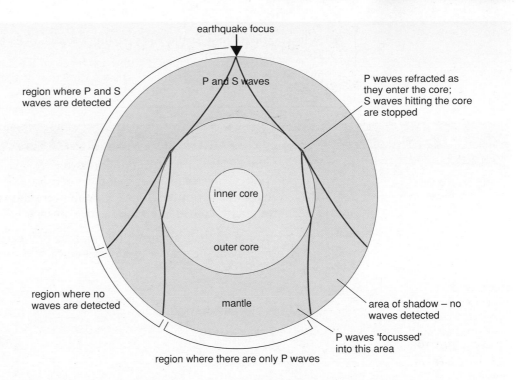

Because the P waves are refracted as they enter the outer core there is a 'shadow' zone where no seismic activity can be detected. It was this fact that enabled scientists to work out how big the Earth's core was and that the Earth had a layered structure.

Did you know?

The centre of the Earth's core is a sphere of solid iron about 3200 km in diameter. Although the temperature is very high, the iron is solid because of the pressure of the materials about it. So great is the pressure that the density of the iron in the core is about 13 g/cm^3 (on the surface of the Earth the density of iron is 7.8 g/cm^3).

Summary

◆ The Earth has a layered structure made up of the **crust, mantle** and **core**.

◆ The **lithosphere** is made up of the crust and the top solid layer of the mantle.

◆ Seismic waves can be longitudinal (**P waves**) or transverse (**S waves**).

Topic questions

1 Earthquakes cause two types of waves.
   a) What are both types of waves called?
   b) What instrument is used to detect them?
2 One type of wave is called a P wave. The other is an S wave.
   a) Which type is transverse?
   b) Which can travel through solids and liquids?
   c) Which can only travel through solids?
   d) Which travels faster?

3 Why do S waves not travel through the core?

4   Why do waves from earthquakes passing through one of the layers in the Earth follow a curved path?

5   Why do waves from earthquakes change direction as they pass from one layer to another?

3.7 Tectonics

Co-ordinated	Modular
10.18	Mod 24 15.5

At one time it was believed that the continents were formed as the once-molten Earth's crust cooled and shrank. For many years most scientists and philosophers also believed that these continents, once formed, were too vast to move so have always been in the same position on the Earth's surface.

However, as early as 1620 an English philosopher, Francis Bacon, realised that there was a great similarity in the shapes of the east coast of South America and the west coast of Africa.

Figure 3.48
Two maps made in 1858 showing how the continents of South America and Africa may once have fitted together

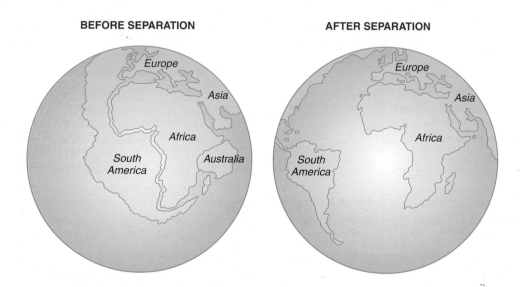

BEFORE SEPARATION   AFTER SEPARATION

In the early 1900s a German scientist, Alfred Wegener, became intrigued about this remarkable fit of the South American and African continents. He was also interested in the similarities in the types of rocks and of plant and animal fossils that were to be found in the coastline areas of continents that were many thousands of miles apart.

In 1912 Wegener developed the idea that continents were moving. He believed that originally there was just one supercontinent, which he called 'Pangaea', and that about 2000 million years ago 'Pangaea' began to break apart (Figure 3.49).

Evidence to support Wegener's ideas have been collected so that today scientists agree with his theory. It is now accepted that:

- The Earth's outermost layer (the **lithosphere**) is made up of a number of large blocks (**tectonic plates**) that are moving relative to each other at speeds of a few centimetres per year. The lithosphere is the more or less rigid outer shell made up from the crust and the uppermost layer of the mantle.

- The plates are moving because of powerful convection currents created within the mantle. These currents are caused by the tremendous heat released by natural radioactive processes taking place in the core of the Earth.

It took some time for scientists to accept Wegeners theory. This is explained later in the chapter.

(a)

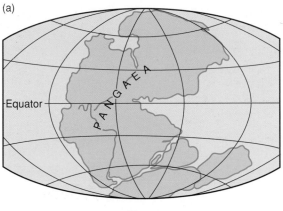

225 million years ago

(b)

200 million years ago

(c)

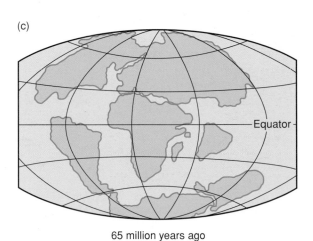

65 million years ago

Figure 3.49
a) The supercontinent of 'Pangaea' 225 million years ago
b) 'Pangaea' breaking up into Laurasia – in the northern hemisphere and Gondwanaland in the southern hemisphere 200 million years ago
c) The land masses 65 million years ago

Earthquakes, volcanoes and the plates

Most earthquakes are caused by stresses set up by movements of the tectonic plates. For this reason most of them occur at the boundaries of the plates in areas where one plate slides past another or where one plate slides under the other.

Magma and gases try to reach the surface of the Earth through weak areas in the lithosphere. Where magma and gases do reach the surface they form volcanoes. Most of the weak areas are found along the boundaries of the tectonic plates.

Figure 3.50
The Earth, showing plate boundaries and direction of movement

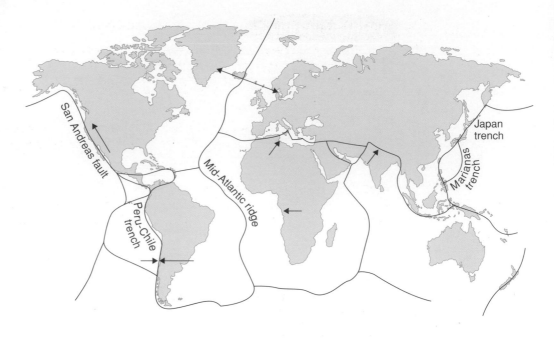

Figure 3.51
Earthquake zones

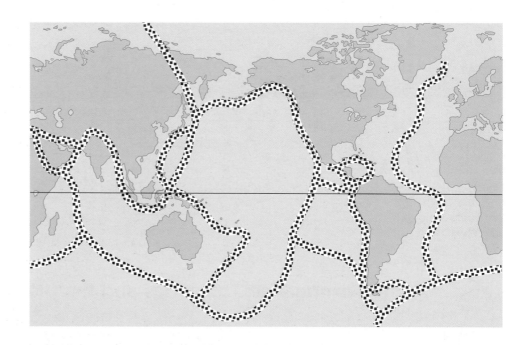

Why the accurate prediction of earthquakes and volcanic activity is difficult

Predicting earthquakes

Although most earthquakes occur in predictable areas scientists cannot give adequate warning of an impending disaster.

Much research has been undertaken to improve the seismic recording equipment and much time has been spent monitoring the many thousands of earthquakes that occur each year. Because earthquakes are caused by stresses building up in rocks below the Earth's surface it is not possible to do much more than monitor the past and present earthquake history of an area and make an intelligent guess as to when the next major earthquake may occur.

Predicting volcanic eruptions

Unlike earthquakes, a volcano will often give a variety of warning signs many years before it erupts. An increase in seismic activity being only one of the signs. Other signs include:

● a change in ground level – as pressure builds up within the magma below the volcano

● an increase in the amount of sulphur dioxide in the area – this indicates that the magma level is rising

● an increase in temperature

● an increase in the flow of magma at the surface.

Even though there are several signs that can be monitored to indicate an increase in volcanic activity, just as with earthquakes a prediction is no more than an intelligent guess.

Why Wegener's theory of Continental Drift was only gradually accepted

Wegener was convinced that all the continents were once joined together as one big 'supercontinent'. He felt that this explained how the continents 'fit' together like a jigsaw, as well as why similar fossils are found in different continents. Another important piece of evidence was that the types of fossils found on one continent suggested that it was once located in a completely different climatic zone. For example, coal deposits (the fossils of tropical plants) in Antarctica led him to believe that this land, now frozen, must once have been much closer to the equator.

However, in 1912 when Wegener published his theory – called Continental Drift – it was not well received even though it seemed to be supported by the scientific data then available. Too many scientists still believed that the continents had been, and still were, permanent features. Part of the reason was because the mechanism of plate tectonics was unknown. Wegener spent the rest of his life trying to find more evidence to support his theory. Wegener died in 1930 and it was not until the 1950s that the exploration of the ocean floor provided evidence which aroused renewed interest in Wegener's theory.

In the 1950s surveys of the ocean floor showed that it contained an enormous mountain range, called the mid-ocean ridge. This range is more than 50 000 km long, sometimes more than 800 km across and in places rises to 4500 m high.

At the same time, other scientists who were using magnetic instruments based on those used to detect submarines, found that there were strange magnetic variations in the rocks on the ocean floor.

Waves

Scientists knew that grains of an iron-rich mineral, called magnetite, are found in volcanic rock and that when the magma from a volcano cools, the grains of magnetite get locked into the rock crystals and line up in the direction of the Earth's magnetic field (see section 1.6). What was surprising about the results was that they showed not only that there were stripes of differently magnetised rocks on either side of the mid-ocean ridge but that the stripes were arranged in the same pattern on either side of the ridge.

Figure 3.52
Magnetic striping on either side of the mid-ocean ridge

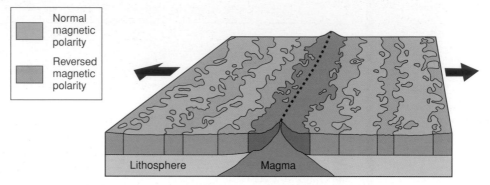

In 1961 scientists began to understand the reasons for the magnetic striping. They considered that the mid-ocean ridge was a weak area in the ocean floor. Along the 50 000 km length of this ridge, magma was erupting and forming new ocean floor. This idea they called sea-floor spreading. They believed that the rocks near the crest of the ridge were the youngest and that these rocks showed the present-day polarity (normal polarity) of the Earth's magnetic field. Stripes of rocks parallel to the crest were older and showed alternate stripes of normal polarity and opposite polarity (reversed polarity) suggesting that the direction of the Earth's magnetic field has been reversed many times during its history.

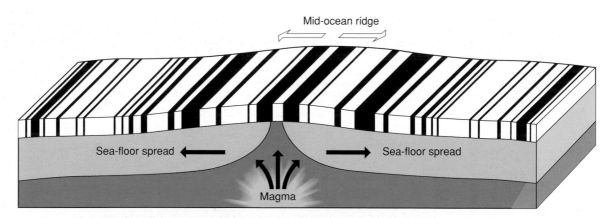

Figure 3.53 ▲
The magnetic striping on the ocean floor

The explanation of the formation of the mid-ocean ridge and the magnetic striping finally convinced scientists that Wegener's ideas had been nearly correct. It is not the continents that move but the plates on which they are fixed. Today the concept of 'Continental Drift' has been replaced with the theory of 'Plate tectonics' .

Movement of the plates

Figure 3.54 shows hot molten magma reaching and bursting through the oceanic crust to form a very long chain of underwater volcanoes – the mid-ocean ridge. The molten magma solidifies to make new crust.

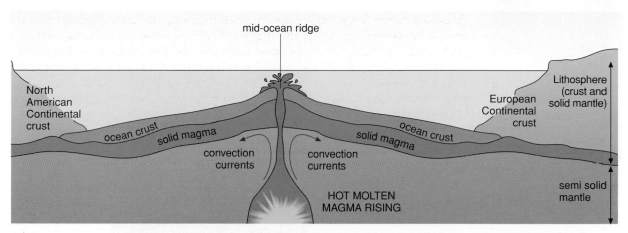

Figure 3.54 ▲
The Atlantic Ocean

Because of the rapid underwater cooling the material of the ocean crust is mainly basalt. The lithosphere (the very top of the mantle and the crust) is a rigid layer around the Earth that is floating on the semi-solid mantle. Convection currents rise through the mantle and reach the crust along the line of the mid-ocean ridge. These convection currents move outwards from the ridge and carry with them the newly formed ocean crust. This effect is called sea-floor spreading.

Because the mid-ocean ridge marks the boundaries between tectonic plates, the formation of new ocean crust and its movement outwards from the ridge means that the continents, such as those of North America and Europe, are moving apart. The Atlantic Ocean is getting wider at a rate of about 4 cm per year.

Figure 3.55 ▼
The Pacific Ocean

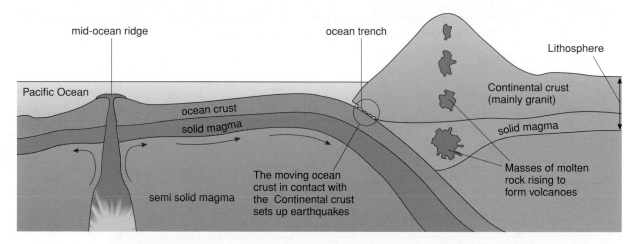

Figure 3.55 shows the ocean crust being carried by convection further and further away from the mid-ocean ridge. As it does so it cools, becoming more dense. Where the ocean crust comes into contact with the continental crust the denser ocean crust gets pushed down by the thicker continental crust. Areas where this happens are called **subduction zones**. The pressures and heat created by the subduction of the ocean crust are sufficient to cause the continental crust to become folded and faulted and the oceanic plate to partially melt. The resulting magma may cause nearby rocks to metamorphose and some of the magma may move up the rock faults to become volcanoes. This is happening along the western side of South America (the Andes).

Did you know?

The Mariana Trench in the Pacific Ocean is caused by a subduction zone. At its deepest it is 11 034 metres below sea level.

In some areas of the world the plates are trying to slide past each other. This is happening in California along the San Andreas Fault.

Figure 3.56
San Andreas Fault

Summary

◆ The lithosphere is cracked into a number of sections called **tectonic plates**.

◆ The tectonic plates move as a result of convection currents in the mantle.

◆ Earthquakes and volcanic activity occur at the plate boundaries.

◆ Wegener proposed a theory to explain the movements of the continents that was not accepted until surveys of the seabed proved that the lithosphere was made up of slowly moving plates.

Topic questions

1 The diagram represents a simplified section through the Earth. Name the parts labelled A, B and C.

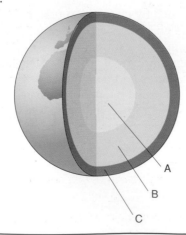

2 What is the lithosphere?

3 What was Pangaea?

4 What are tectonic plates?

5 What causes tectonic plates to move?

6 What causes a volcano?

7 What causes an earthquake?

8 How do we know that the Earth's magnetic field has varied many times during the Earth's history?

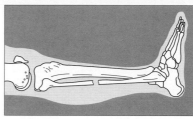

Examination questions

1 The boxes on the left show some types of electomagnetic radiation. The boxes on the right show some uses of electromagnetic radiation.

Copy the diagram and draw a straight line from each type of radiation to its use. The first one has been done for you.

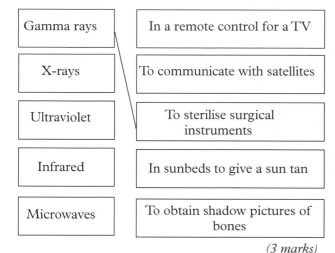

(3 marks)

2 a) The diagram represents the electromagnetic spectrum. Four of the waves have not been named. Copy the diagram and draw lines to join each of the waves to its correct position in the electromagnetic spectrum. One has been done for you.

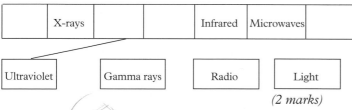

(2 marks)

b) Copy and complete the following sentence by choosing the correct answer from the three lines in the box.

The speed of radio waves through a vacuum is

faster than
the same as	the speed of light through a vacuum.
slower than

(1 mark)

c) i) Before sunbathing it's a good idea to apply a sun cream to your exposed skin. Why?
(1 mark)

ii) From which type of electromagnetic wave is sun cream designed to protect the skin?
(1 mark)

d) The diagram shows an X-ray photograph of a broken leg.
Bones show up white on the photographic film. Explain why. *(2 marks)*

3 a) The diagram shows an electric bell inside a glass jar. The bell can be heard ringing.

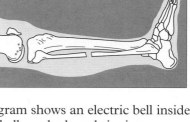

Copy and complete the following sentences, by choosing the correct line in each box.

When all the air has been taken out of the glass jar,

the ringing sound will
stop.
get louder.
get quieter.

This is because sound
travels faster
travels slower	through
cannot travel
a vacuum. *(2 marks)*

b) The microphone and cathode ray oscilloscope are used to show the sound wave pattern of a musical instrument.

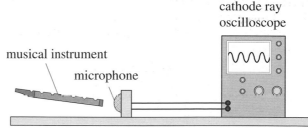

One of the following statements describes what a microphone does. Identify the correct statement.
(1 mark)

● A microphone transfers sound energy to light energy.

● A microphone transfers sound energy to electrical energy.

● A microphone transfers electrical energy to sound energy.

135

c) Four different sound wave patterns are shown. They are all drawn to the same scale.

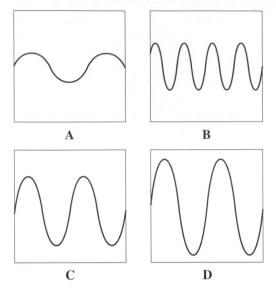

i) Which sound wave pattern has the highest pitch? Give a reason for your answer.
(2 marks)

ii) Which sound wave pattern is the loudest? Give a reason for your answer. *(2 marks)*

d) i) The frequency of some sounds is too high for humans to hear. Which of the following words describes this sound.

microwave   ultrasound   ultraviolet

(1 mark)

ii) Give **one** use for this type of sound wave.
(1 mark)

4 a) The student is using a microphone connected to a cathode ray oscilloscope (CRO).

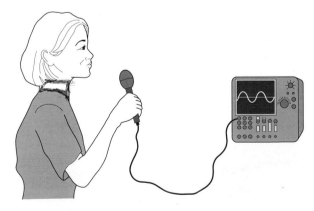

The CRO displays the sound waves as waves on its screen. What does the microphone do?
(2 marks)

b) The amplitude, the frequency and the wavelength of a sound wave can each be either increased or decreased.
i) What change, or changes, would make the sound quieter? *(1 mark)*
ii) What change, or changes, would make the sound higher in pitch? *(1 mark)*

c) People can generally hear sounds in the frequency range 20 Hz to 20 000 Hz.
i) What are very high frequency, and inaudible, sounds with frequencies **greater** than 20 000 Hz called? *(1 mark)*
ii) Give **two** uses for very high frequency sounds. *(2 marks)*

d) The diagram shows sound waves approaching a gap in a wall.

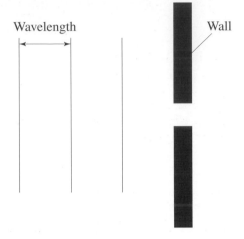

i) Copy and complete the diagram to show what will happen to the sound waves on the other side of the wall. *(2 marks)*
ii) What is the name of this effect? *(1 mark)*
iii) What would the width of the gap need to be for this effect to be most pronounced? *(1 mark)*

5 The diagram represents the structure of the Earth.

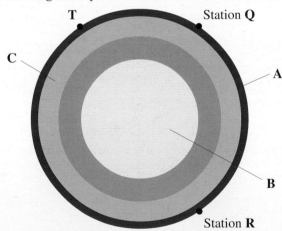

a) On the diagram, name the parts **A**, **B** and **C**.
b) An earthquake occurs at the point **T** on the Earth's surface. Two types of shock wave are produced by the earthquake, P waves and S waves.

Describe **two** similarities and **two** differences between P waves and S waves as they travel through the Earth. *(4 marks)*

c) State whether P waves or S waves or both will reach:
i) Station Q *(1 mark)*
ii) Station R. *(1 mark)*

6 The diagrams below show some pieces of glass.

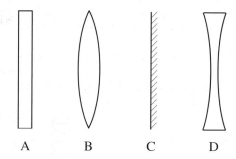

A B C D

a) Which of **A, B, C** and **D** is
   i)   a converging lens?
   ii) a diverging lens? *(2 marks)*
b) Copy and complete the diagram below to show what happens to the rays of light when they pass through **B**.

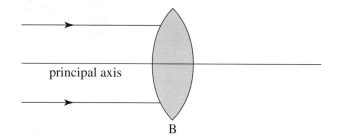

principal axis

B

(4 marks)

7 a) An object OB is placed 12 cm in front of a converging lens of focal length 9 cm. The diagram below is drawn to scale.

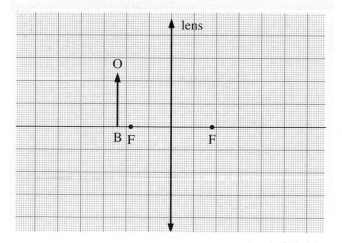

i)   Draw the ray diagram on graph paper to show the position and size of the image. Draw and label the image.
ii) Write down two ways in which the image is different from the object.
   (6 marks)
b) Cameras use converging lenses to produce an image of an object. Give two ways in which the image produced on the film is different from the object. *(2 marks)*

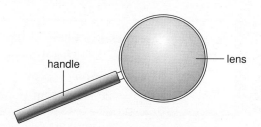

handle   lens

8

When some people are reading a book with very small print, they may use a lens like the one shown in the diagram.
a) State the type of lens used.
b) Explain, in as much detail as you can, how the lens makes it easier to read the print.
   (4 marks)

Chapter 4
The Earth and beyond

Key terms artificial satellite · Big Bang · black hole · comet · fusion · galaxy · geostationary satellite · gravity · light year · Milky Way · moon · orbit · planet · red giant · red shift · satellite · solar system · star · Sun · Universe · white dwarf

4.1

Co-ordinated	Modular
10.19	Mod 11
	12.4

The solar system

Within our **solar system** there is one **star** (the **Sun**) and nine **planets**. Each of the nine planets follows a regular path, called an **orbit**, around the Sun. Mercury and Venus are the only planets which do not have a **moon**.

Also in orbit around the Sun are **comets**.

Figure 4.1
Parts of the solar system

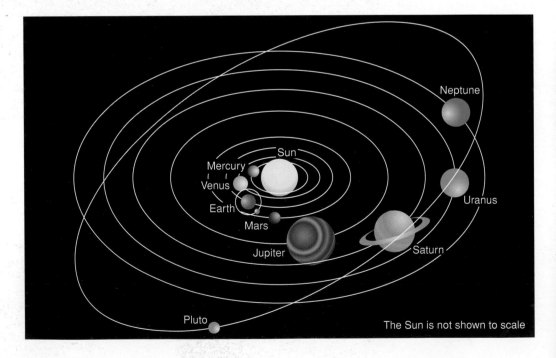

The Sun is not shown to scale

The planets

Each planet travels on a path or orbit around the Sun. Some planets have an orbit which is almost a circle, with the Sun at the centre of the circle.

Other planets have orbits like squashed circles, these are elliptical orbits. The orbit of Pluto is so elliptical that at times it is closer to the Sun than Neptune.

Figure 4.2

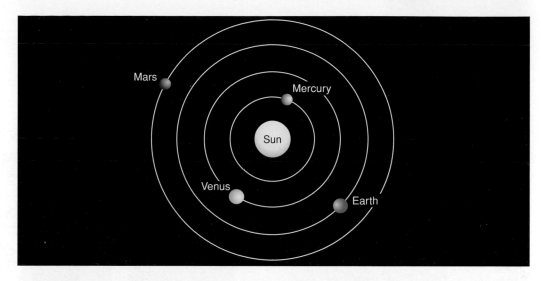

Except for Pluto, all the planets orbit the Sun in the same plane.

Figure 4.3

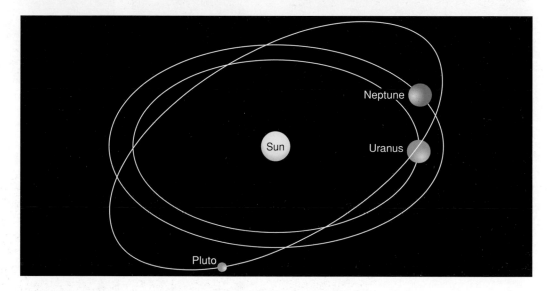

The Moon

The Moon is the Earth's natural **satellite**. Like the Earth orbits the Sun, so the Moon orbits the Earth. One orbit of the Moon takes 27 days or 1 lunar month. Like the planets, the Moon is non-luminous. We see the Moon when it reflects light from the Sun towards us.

Figure 4.4

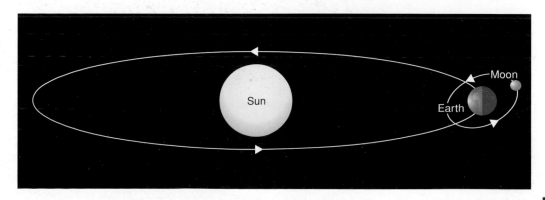

Many of the other planets are orbited by one or more moons. Saturn has 20 moons.

Comets

Comets are thought to be chunks of frozen rock covered by huge amounts of frozen water and gases. Comets also orbit the Sun, however compared to the planets, their paths are more elongated and in different planes. Comets can only be seen for the short amount of time that their orbit passes close to the Sun. At this time energy from the Sun causes some of the frozen gases and water to vaporise, giving the comet its spectacular tail.

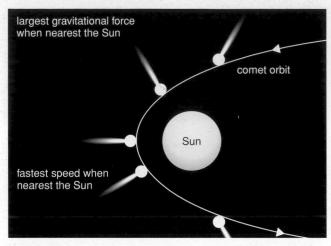

Figure 4.5
The path of a comet

Figure 4.6
The comet Hyakutake showing its glowing tail

Figure 4.7 ▶
Small masses, small force of attraction

Gravitational forces in the solar system

Gravity is a force which tries to pull objects together. It is a force of attraction. Gravity acts between all objects, no matter how big or how small. The bigger the mass of the objects, the bigger the gravity force. The force is only big enough to feel if one of the objects has a very large mass, such as one of the planets or a star.

The gravitational pull of the Sun affects all objects in the solar system. It is the gravitational force between the Sun and a planet which keeps the planet in its orbit around the Sun. If the Sun's gravitational pull suddenly stopped pulling, the Earth and all the other planets would shoot off into space.

Figure 4.8
Large masses, large force of attraction

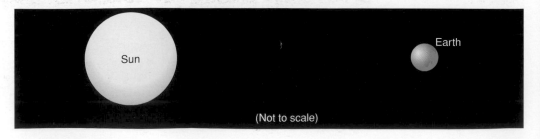

(Not to scale)

The orbit of a moon around a planet is due to the gravitational pull between the moon and the planet.

As the distance between objects in the solar system increases, so the gravitational force between them decreases. The force of gravity between planets is weak because of the great distances between the planets.

Figure 4.9

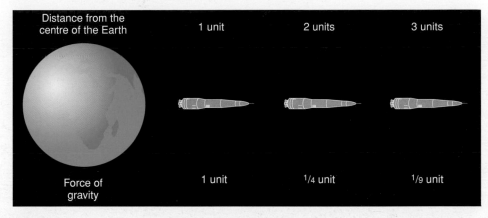

Did you know?

The spacecraft Cassini, launched in 1997, will reach Saturn in the year 2004. A probe will then explore the atmosphere and surface of Titan, Saturn's largest moon.

Figure 4.9 shows how the force of gravity between two masses changes with distance. If the distance between the two masses doubles, the force of gravity goes down by a factor of four. If the distance between the two masses trebles then the force of gravity goes down by a factor of nine. This is called the inverse square relationship.

Gravity can be used to change the direction of a spacecraft. Figure 4.10 shows the path taken by the Cassini spacecraft. Each time the spacecraft approaches a planet, the gravitational force will swing and accelerate it for the next part of its journey.

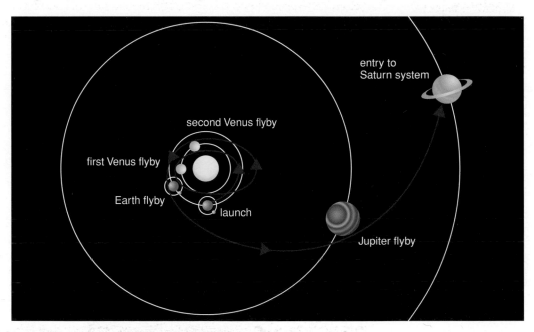

Figure 4.10

Artificial satellites

These are held in orbit by the gravitational pull of the Earth and the speed at which they are moving. The first **artificial satellite**, Sputnik 1, was put into orbit in 1957. Since then, hundreds of satellites have been launched into space. Artificial satellites have many different uses.

The Earth and beyond

1 Observation of the Earth

Satellites used to observe the Earth are usually put into a low polar orbit. Passing over the Earth they scan the surface sending back detailed pictures. Such things as volcanic activity, the position of an oil slick or the path of a hurricane can all be watched and monitored. Because the Earth spins, different parts of the Earth are seen on each rotation.

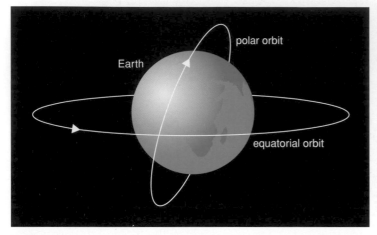

Figure 4.11

Figure 4.12
A satellite picture of the River Thames in London

2 Weather monitoring

Satellites, such as the European Space Agency's Meteostat, are put into high equatorial orbits. They monitor the changing weather patterns over one part of the Earth's surface. With this information, weather forecasts can be made more accurate.

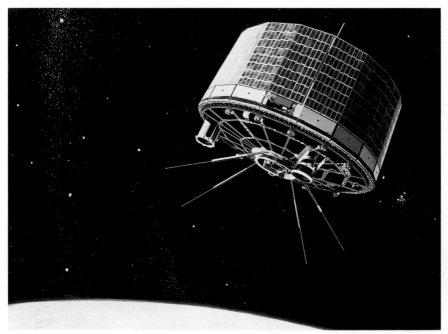

Figure 4.13
Meteostat in orbit

Figure 4.14
The weather pattern seen from a satellite

3 Exploration of the solar system

Satellites can produce sharp images of the solar system. This is because they do not have any atmospheric interference. Space probes can also travel to other planets, take photographs and send the images back to Earth.

Figure 4.15
A satellite transmission system for television

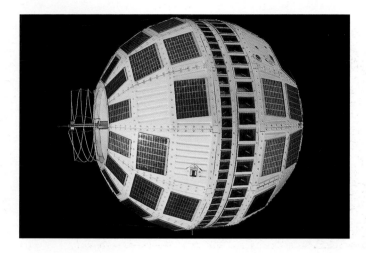

Figure 4.16
Satellite dishes receive the TV signal from the satellite

4 Communication systems

Geostationary satellites used in communication systems are put into high equatorial orbits. They travel at a speed which takes them once around the Earth every 24 hours. This means that the satellite moves around the Earth at the same rate as the Earth spins.

In this way the satellite stays above the same point on the Earth's surface. These satellites can be used to send television programmes and telephone messages around the world.

Television programmes are transmitted to the satellite using microwaves. The satellite then transmits the programmes back to an area of the Earth as large as Europe. A dish aerial fixed in the correct direction can then be used to pick up the programme.

Summary

◆ The **orbits** of the inner planets around the **Sun** are more or less circular; those of the outer planets are elliptical.

◆ **Comets** have very elliptical orbits.

◆ All bodies attract each other with a gravitational force.

◆ Communication and monitoring satellites are put into geostationary orbit.

Topic questions

1. Using words from the box copy and complete the following sentences. Each word may be used once or not at all.

galaxy	moons	star	universe

   The solar system has one _____ called the Sun. There are nine planets in orbit around the Sun. Some planets have one or more _____ in orbit around them.

2. Copy and complete the following sentences.

   Planets orbit a _____ .

   A moon orbits a _____ .

3 The box contains the names of eight of the nine planets in the solar system.

| Earth Jupiter Mars Mercury Neptune Pluto Saturn Uranus |

  a) Name the planet which has not got its name in the box.
  b) Which planet has the shortest orbit?
  c) Name the force which keeps a planet in its orbit.

4 The diagram below, which is not drawn to scale, shows a communications satellite in orbit above the Earth.

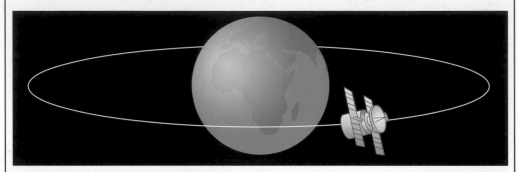

The satellite is in a geostationary orbit. Explain what is special about this sort of orbit.

5 The diagram below, which is not drawn to scale, shows the path of one kind of object in the solar system. The object is not a planet or moon.

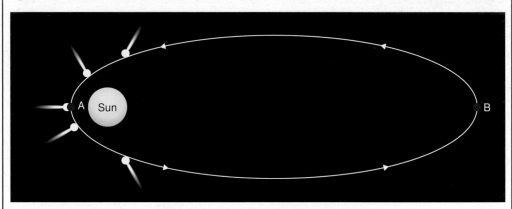

  a) What is the name of this kind of object?
  b) What is the force which causes the object to change direction?
  c) Why is the force greater at point A than at point B?
  d) At which point A or B will the object be moving the fastest?

6 Saturn has about 100 times the mass of the Earth, and Saturn is about 10 times further away from the Sun than the Earth. How does the Sun's gravitational pull on Saturn compare to the Sun's gravitational pull on the Earth?

4.2   The wider Universe

Co-ordinated	Modular
10.20	Mod 11
	12.5

A **galaxy** is a vast number of stars held together by the force of gravity. The Sun is just one of the 100 000 million stars which make up the **Milky Way** galaxy. Figure 4.17 shows what the Milky Way would look like if it were being seen from above and from the side. The whole galaxy is so huge that it takes light 100 000 years to travel from one side to the other.

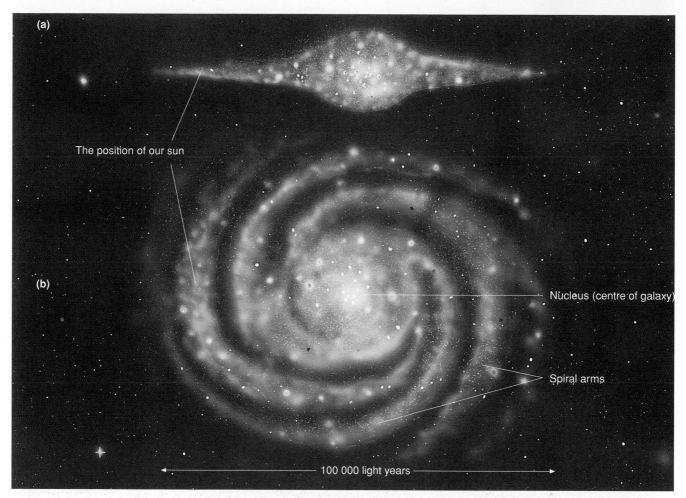

(a)

The position of our sun

(b)

Nucleus (centre of galaxy)

Spiral arms

100 000 light years

Figure 4.17 ▲
The Milky Way galaxy,
a) viewed from the side,
b) viewed from above

Figure 4.18 ▶
The Andromeda galaxy

The Milky Way is not the only galaxy. Using large telescopes, millions of similar galaxies can be seen in all directions in space. The **Universe** may contain billions (thousands of millions) of galaxies, each having about as many stars as the Milky Way.

The Earth and beyond

The distance between stars and between galaxies is so big that it is not sensible to measure it in kilometres. The distances are measured by saying how long it takes light from the star or galaxy to reach us. These distances are called **light years**. A light year is the distance that light travels in one year.

Since the speed of light is 300 000 km/s, one light year is roughly 9 500 000 000 000 kilometres.

Evidence for the origin of the Universe

By looking at light sent out by distant galaxies, scientists have come to the conclusion that the Universe is expanding.

In 1929 an American astronomer, Edwin Hubble, examined the spectrum of the light from various galaxies. He noticed that whichever galaxy he looked at, the spectra were composed of light of longer wavelengths, that is, the spectra (spectral lines) were always shifted towards the red end. He called this effect **red shift**. He also noticed that spectra for the closer galaxies showed less red shift than did the spectra for galaxies further away. Scientists have discovered that this shift in the wavelength of light to the red end of the spectrum occurs when a light source is moving away. The faster the light source is moving away, the greater the red shift. Therefore Hubble concluded that all galaxies were moving away from Earth and that the farther the galaxy the faster it was moving away.

In 1948 George Gamow, a Russian physicist, proposed that if all the galaxies are moving away at high speed then there must have been a time when they were all concentrated in a single place. From this comes the idea that the Universe was then created by a massive explosion. This created the dust and gases which went on to form the planets and the stars. He used the term **Big Bang** to describe that moment when all the matter that was concentrated in one place exploded and started to expand outwards. The Universe has been expanding ever since.

Formation of stars

Stars do not last for ever. They go through a life cycle from birth to death. Like all stars our Sun was formed from a huge cloud of mainly hydrogen gas. Over millions of years this cloud was pulled inwards and condensed into a smaller volume by the force of gravity.

Figure 4.19
A cooling mass of gas in the great Nebula in Orion

This contraction caused reactions which produced vast amounts of energy. This energy was released as heat and light causing a huge temperature rise. At this stage the star was born and further contraction stopped.

During the main period of its life, a star remains in a stable state. The inward gravitational forces trying to make it collapse are balanced by the enormous outward pressure created by the high temperature at the centre of the star. This period of stability is likely to last for billions of years. Our Sun is in its stable period and about half way through its life.

The reaction which produces the vast amounts of energy is called nuclear **fusion**. In this process hydrogen nuclei fuse together to form helium nuclei. This process continues during the stable period of a star, allowing it to radiate heat and light energy. As the hydrogen is used up, the mass of the star decreases, causing the next phase in the life cycle of the star – the formation of a red giant.

Figure 4.20
A stable star

Did you know?

The sun is converting 4 million tonnes of its mass into energy every second. It has been doing this for the last 5000 million years and will continue for another 5000 million years.

Eventually the reactions that were responsible for the release of the vast amounts of energy begin to stop and the mass of the star decreases. The inward gravitational forces are no longer able to balance the outward pressure created by the high temperatures. The star gradually expands and cools to become a **red giant**. When our Sun reaches this stage it will be so large that the Earth will end up inside it!

What happens in the final stages of the life cycle of a star depends on its size. Gravitational forces cause the star to contract, causing another rise in temperature and further energy release. The star is now a **white dwarf** and the matter from which it is made may be many millions of times more dense than any matter on Earth.

Figure 4.21
A supernova

The Earth and beyond

If a white dwarf is formed from a small star, like the Sun, it will fade and become a black dwarf.

However if a star is massive enough, then the final stage after becoming a very dense white dwarf is for the star to explode as a supernova, scattering gas and dust into space. During a supernova stage the dying star emits more energy into space in a few seconds than is produced by the Sun in millions of years.

The matter left behind from the exploding white dwarf may form a very dense neutron star.

Did you know?

Neutron stars are thought to have very strong magnetic fields which cause splits in the surface of the star allowing it to emit vast amounts of gamma radiation.

Neutron stars are so dense that if the ball on the end of a ball-point pen were made of matter packed as densely, the ball would have a mass of about 91 000 tonnes.

Figure 4.22
The life cycle of a star

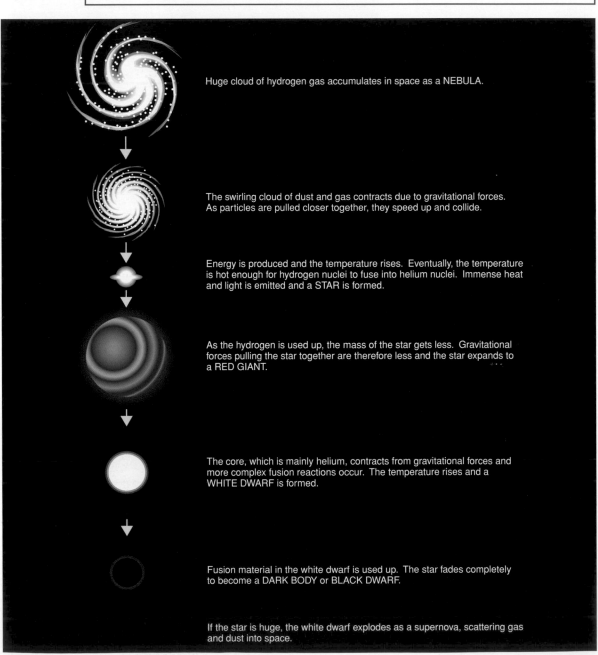

Huge cloud of hydrogen gas accumulates in space as a NEBULA.

The swirling cloud of dust and gas contracts due to gravitational forces. As particles are pulled closer together, they speed up and collide.

Energy is produced and the temperature rises. Eventually, the temperature is hot enough for hydrogen nuclei to fuse into helium nuclei. Immense heat and light is emitted and a STAR is formed.

As the hydrogen is used up, the mass of the star gets less. Gravitational forces pulling the star together are therefore less and the star expands to a RED GIANT.

The core, which is mainly helium, contracts from gravitational forces and more complex fusion reactions occur. The temperature rises and a WHITE DWARF is formed.

Fusion material in the white dwarf is used up. The star fades completely to become a DARK BODY or BLACK DWARF.

If the star is huge, the white dwarf explodes as a supernova, scattering gas and dust into space.

The red giant reaches its maximum size when all the hydrogen has been converted to helium. The temperature at the centre of a red giant causes the helium nuclei to fuse together to form heavier elements. More elements form as the star contracts to become a white dwarf. During a supernova the nuclei of elements heavier than hydrogen are sent into space where they can form part of the contents of future stars. Future stars start life with a richer supply of heavier elements that earlier generations of stars.

Nuclei of the heaviest elements are present in the Sun and atoms of these elements are present in the inner planets of the solar system. This suggests that the solar system was formed from the material produced when earlier stars exploded.

Black holes

Sometimes after a supernova the matter that is left behind and not scattered into space is so dense and the gravitational forces so strong that the matter collapses to form a black hole.

Black holes have been given this name because light can enter but not escape due to the tremendous gravitational forces associated with them. We cannot see black holes but we can observe their effects on their surroundings. For example, we can detect the X-rays emitted when gases from a nearby star spiral into a black hole.

Did you know?

Even the Earth could be made into a black hole if it could be squashed down to the size of tennis ball.

The search for life elsewhere in the galaxy
What should we be looking for?

So far, in the quest for evidence of life beyond our planet the search has been restricted to the sending of manned and unmanned lunar modules to the Moon, unmanned probes to Mars and Venus.

Figure 4.23
An image of the surface of Mars taken by the Mars Pathfinder space probe

Scientists have assumed that if life exists or has existed, any evidence remaining will be based on the life processes associated with living things on Earth. Because they assume that water must be available to support photosynthesis and that oxygen must be available to support respiration, scientists have designed and set up experiments on the surfaces of both the Moon and Mars to test samples of soil and rock.

The Earth and beyond

These experiments were designed:

- to test for the presence of water and oxygen.

- to measure whether there were any changes in the gases that might indicate that photosynthesis and respiration were taking place.

- to examine soil samples to discover if living organisms were present or whether there was fossil evidence that living organisms had once existed.

It is important to realise that our atmosphere would contain little or no oxygen if it were not for the existence of plants.

Much time and money has been spent but so far there has been no success. However, it may be that they have been looking for the wrong evidence, for recently on Earth some organisms have been discovered surviving in conditions that were once thought impossible.

- Microbes have been found in the sulphur-laden atmosphere of volcanoes, hot springs and geysers.

- Deep-diving submarines have found microbes that get their energy from the heat produced by hot rocks.

- Other microbes have been found deep underground that use hydrogen as their energy source.

Did you know?

In 1984 a meteorite was discovered in Antarctica that scientists are convinced came from Mars thousands of years ago. Examination of the meteorite seemed to show the presence of microscopic worm-like structures in the rock.

The Mars meteorite

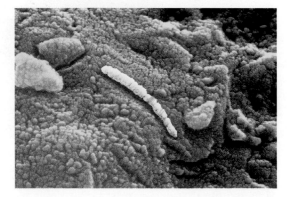

Are these worm-like structures fossilised Martian bacteria?

Should we be looking for intelligent life?

In 1960 an American called Frank Drake was the first scientist to start a careful search for intelligent radio signals from space. He spent six hours each day for about four months using a 25 m radio telescope. He was unsuccessful.

Drake's lack of success has not stopped radio astronomers from trying to search for extraterrestrial life in space (SETI). However, 40 years of SETI have still failed to find anything, even though there have been tremendous advancements in the technology available.

Figure 4.24
Frank Drake and Jill Tarter lead the SETI team. They stand beneath the radio telescope they use to look for radio signals in space

What do we do now?

It is now realised that to send manned craft to other planets and moons is costly and dangerous. For the exploration of deep space a manned craft will take too long. So it seems that any further exploration of space will be undertaken using unmanned craft and robots.

Because it is now known that on Earth there are organisms that can survive in extremely inhospitable conditions, scientists are beginning to reconsider looking for life in what they once considered to be unsuitable areas of the solar system. For example, photographs from the 1979 *Voyager* spacecraft mission to Jupiter and its neighbours showed that the surface of Europa, one of Jupiter's moons, seems to be covered in ice. Further photographic evidence from the recent *Galileo* mission to Jupiter shows that the surface of Europa could be an icy crust between 9–16 km thick. There is some evidence to show that the interior of this moon may just be warm enough to have melted some of the ice and that liquid water may be present below the ice. The Americans are considering sending a lander to Europa with the aim of drilling through the ice and sending a probe called a hydrorobot to investigate the contents of the water.

Figure 4.25
The hydrorobot

Summary

◆ The **Universe** is made up of a very large number of **galaxies**.

◆ **Stars** are formed when dust and gas are pulled together.

◆ Stars go through a life cycle from birth to death.

◆ So far, no life seems to have been found anywhere else in the Universe.

◆ The discovery of the concept of the **red shift** led to the idea of the **Big Bang** as being the origin of the Universe.

Topic questions

1 Which one of the following quantities can be measured in light years?

    distance speed time

2 The solar system is part of which galaxy?

3 Rewrite the following in order of size. Start with the smallest.

    galaxy planet solar system star Universe

4 What is a galaxy?

5 The Andromeda Galaxy is about 2.5 million light years away. Why are humans unlikely to ever explore it?

6 Explain how the 'Big Bang' theory accounts for the creation of the Universe.

7 What does the 'Big Bang' theory predict is happening to the size of the Universe?

8 Explain how stars are formed.

9 By what process does a star get its energy?

10 For the millions of years between its 'birth' and its 'death' a star is stable. Describe the two main forces at work in the star during this time.

11 Describe the process by which a red giant changes into a white dwarf.

Examination questions

1 a) Copy and complete each sentence by choosing the correct word or phrase from the box. Each word or phrase should be used once or not at all.

> milky way   moon   planet   solar system   star   universe

The Sun is the nearest _____ to the Earth.
The Sun is in the galaxy called the _____.
Within the _____ there are millions of galaxies.
Pluto is orbited by one _____. *(4 marks)*

b) The diagram shows the path taken by the Voyager 2 spacecraft.

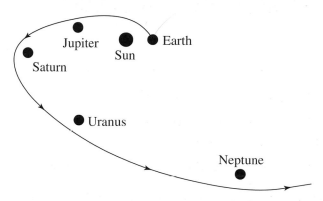

Choosing from the forces in the box, which force caused the spacecraft to change direction each time it got close to a planet?

> air resistance   friction   gravity

(1 mark)

2 a) The table gives some information about four planets.

Planet	Average distance from the Sun in million km	Average time to complete one orbit in Earth years	Average orbital speed in km/sec
Jupiter	800	12	13.0
Saturn	1400	30	9.6
Neptune	4500	165	5.2
Pluto	5900	248	4.7

i) Draw a graph of each planet's average orbital speed against the distance the planet is from the Sun. Plot distance from the Sun on the horizontal axis and orbital speed on the vertical axis.

(3 marks)

ii) How does the average orbital speed of a planet vary with its average distance from the Sun? *(1 mark)*

iii) The average distance between Uranus and the Sun is 2900 million kilometres. Use the graph to predict the average orbital speed of Uranus. *(1 mark)*

b) The diagram shows the position of Saturn in July 1984 and July 1986.

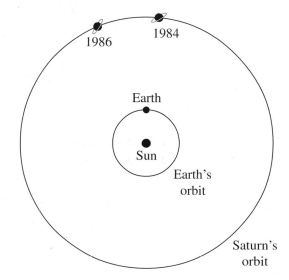

i) Saturn takes 30 Earth years to complete one orbit of the Sun. Copy the diagram and mark the position of Saturn in the year 2000. *(1 mark)*

ii) Suggest why it was difficult to see Saturn in July 2000. *(1 mark)*

3 A satellite in a stable Earth orbit moves at constant speed in a circle, because a single force acts on it.

a) i) Name the force acting on the satellite. *(1 mark)*

ii) State the direction of this force. *(1 mark)*

b) Communications satellites and satellites used to observe the Earth are placed in different orbits.

i) Describe the orbit of a communications satellite. *(3 marks)*

ii) Describe the orbit of a satellite used to observe the Earth. *(2 marks)*

iii) Explain why the satellites are placed in different types of orbit. *(3 marks)*

c) Explain, in terms of its orbit, why a comet is rarely seen from Earth. *(2 marks)*

4 a) The Cassini spacecraft launched in 1997 will take seven years to reach Saturn. The journey will take the spacecraft close to several other planets.

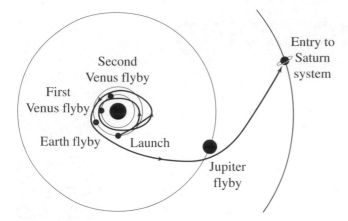

Each time the spacecraft approaches a planet it changes direction and gains kinetic energy. Explain why. *(2 marks)*

b) The Big Bang theory attempts to explain the origin of the Universe.
   i)   What is the Big Bang theory?   *(1 mark)*
   ii) What can be predicted from the Big Bang theory about the size of the Universe?   *(1 mark)*

c) i)   Explain how stars like the Sun were formed.   *(2 marks)*
   ii) The sun is made mostly of hydrogen. Eventually the hydrogen will be used up and the Sun will 'die'.
       Describe what will happen to the Sun from the time the hydrogen is used up until the Sun 'dies'.   *(3 marks)*

Chapter 5
Using energy and doing work

Key terms conduction • convection • efficiency • elastic potential energy • electrical energy • fossil fuels • free electrons • generator • geothermal energy • global warming • gravitational potential energy • greenhouse effect • hydroelectric • kinetic energy • non-renewable resources • power • radiation • renewable energy • turbine • work

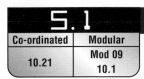

5.1	
Co-ordinated	Modular
10.21	Mod 09 10.1

Thermal energy transfers

Radiation

Radiation is the energy transfer that takes place without the movement of any particles. Indeed energy transfer by radiation is most efficient when no particles are present. It is the process by which our planet receives heat energy from the Sun through the vacuum of space.

All objects radiate energy. The hotter the object the more heat energy it radiates.

The heat energy radiated is called infrared radiation. Infrared radiation is one of the members of the family of waves called the electromagnetic spectrum (see section 3.3).

Giving out radiation (radiation emission)

- Hot tea in a light-coloured teapot will transfer energy to the air slower than tea in a dark-coloured teapot.

- Hot tea in a highly-polished shiny teapot will transfer energy to the air slower than tea in a non-shiny teapot. (A non-shiny surface is described as having a matt finish.)

So, hot surfaces that are light coloured and shiny will transfer infrared radiation to the air slower than hot, dark-coloured matt surfaces.

Taking in radiation (radiation absorption)

The outside metal of a dark-coloured car warms up quicker in the Sun than does the metal on a light coloured car. A polished shiny car will not get as hot as an unpolished car.

This happens because:

- dark-coloured cold surfaces absorb infrared radiation more quickly than do light-coloured cold surfaces

- shiny cold surfaces reflect more infrared radiation than do matt cold surfaces.

Figure 5.1

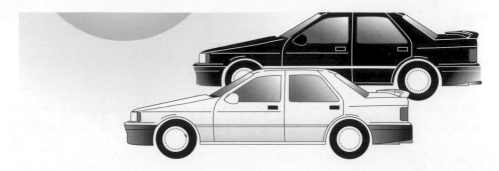

Figure 5.2
A summary of radiation

Good absorbers of infrared radiation (if cold these get warm quickly)	Good emitters of infrared radiation (if hot these cool down quickly)	Poor absorbers of infrared radiation (if cold these get warm slowly)	Poor emitters of infrared radiation (if hot these cool down slowly)
dark-coloured surfaces	dark-coloured surfaces	light-coloured surfaces	light-coloured surfaces
non-shiny (matt) surfaces	non-shiny (matt) surfaces	shiny surfaces	shiny surfaces

Infrared radiation and the atmosphere

The carbon dioxide in the atmosphere acts like a blanket around the Earth to keep the Earth warm. Not all of the infrared radiation coming from the Sun and hitting the Earth's surface is absorbed by the Earth's surface. Some of it is reflected back into space.

Figure 5.3
Some of the Sun's radiation is reflected back into space

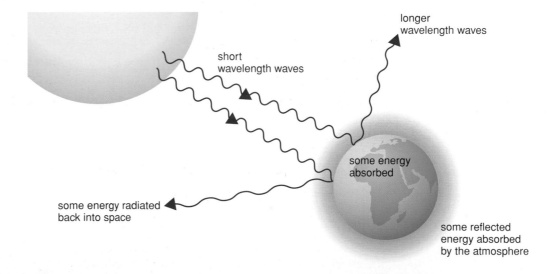

longer wavelength waves

short wavelength waves

some energy absorbed

some energy radiated back into space

some reflected energy absorbed by the atmosphere

The carbon dioxide in the atmosphere absorbs some of this reflected infrared radiation so keeping the Earth reasonably warm. This effect is known as the **greenhouse effect**. As more and more **fossil fuels** are burned, more carbon dioxide passes into the atmosphere. This has the effect of absorbing more reflected infrared radiation so making the atmosphere even warmer. This warming is called **global warming**.

Conduction

Conduction is the process by which energy transfers take place in solids. Metals are better conductors than non-metals. If one end of a metal bar is heated in a Bunsen flame, the heat energy from the flame is quickly transferred along the bar from the hot end to the cold end. This happens because metals contain particles that can move, called **free electrons**, and particles called ions (ions are atoms that have lost an electron) that can only vibrate (Figure 5.4). At the heated end of the bar, the heat energy in the flame:

- increases the rate at which the closely packed ions vibrate. These vibrating ions collide with their neighbours. These collisions give the neighbouring ions more energy so they vibrate more rapidly.

- increases the energy of the free electrons. These free electrons travel long distances very rapidly between collisions and so transfer energy along the bar quickly.

Figure 5.4
Free electrons between the ions in a metal

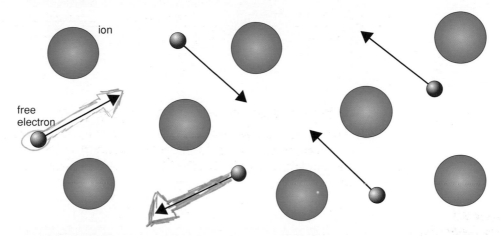

This happens all along the bar and so the heat energy is transferred from the heated end along the bar by a series of collisions between neighbouring ions and by the movement of the free electrons.

Non-metals have no free electrons so the energy transfer process is much slower.

Convection

Convection is a form of energy transfer that takes place in liquids and gases (fluids).

Figure 5.5 shows water being heated.

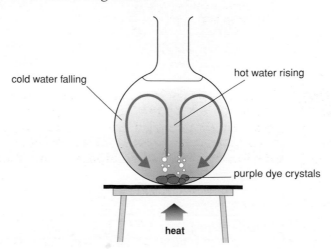

Figure 5.5
Energy transfer in a liquid

As the water at the bottom is warmed up, the particles gain heat energy. This extra energy causes:

- the particles to vibrate with a bigger amplitude
- the particles to take up more space
- the water in the warmer region to expand
- the warm water as it expands to become lighter (less dense) than the cooler water around it
- the warm water to rise.

As the warm water rises, cooler water flows in to replace it. This water gets heated and rises. The movements of hot and cold water due to the changes in density are called convection currents. Convection currents stop when all parts of the water are at the same temperature. Convection currents occur because cold, dense water sinks and warm less dense water rises.

Did you know?

The Atlantic Gulf Stream is caused by convection currents.

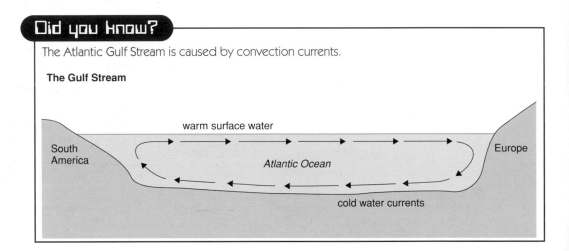

The Gulf Stream

Ways to reduce heat loss

Figure 5.6

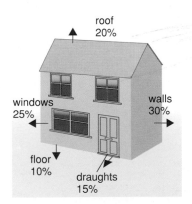

Different amounts of heat are lost through the roof, the walls, the windows, the floor and the doors of a home. These are shown in Figure 5.6.

There are many devices that have been designed to reduce unwanted heat loss from a home. Information about some of these devices is given in Figure 5.7 and the cost of these are given below.

It could take up to 40 years if the money saved in heating bills by having double glazing fitted was used to pay for the fitting.

Ways of keeping the house warm	Pay-back time using money saved on the heating bills
Cavity wall insulation	3 years
Loft insulation	1.5 years
Double glazing	40 years

Although double glazing is very expensive it does act as a sound insulator.

Figure 5.7

Device	How heat energy losses are reduced
Cavity wall insulation convection currents in cavity / foam – no convection currents possible	The cavity between two outside walls is filled either with mineral wool or foam. • the foam and mineral wool both trap a large number of tiny air pockets • Because the air is trapped it cannot move, so convection currents cannot occur. • The still air is also a good insulator.
Loft insulation	The fibres in the mineral wool trap a large number of tiny air pockets. • Because the air is trapped it cannot move, so convection currents cannot occur. • The still air is also a good insulator.
Double glazing	• Thick glass is used so reducing heat loss by conduction. • The still air trapped between the two panes reduces heat loss by convection and conduction.
Thick curtains	• These stop cold air blowing into a room. • They trap air between the wall and the curtain so reducing heat loss by conduction.

Summary

◆ **Radiation** is the transfer of energy by infrared radiation.

◆ Radiation can pass through a vacuum and can be reflected.

◆ **Conduction** of heat energy is the result of the movement of free electrons and collisions between ions.

◆ Metals are good conductors of heat energy because they have a lot of free electrons.

◆ Non-metallic substances, including trapped air, are poor conductors of heat energy.

◆ **Convection** is the movement of more energetic particles in liquids and gases.

Topic questions

1 Explain why metals are good heat conductors.

2 The elements in an electric kettle are always found near the bottom of the kettle. Explain why.

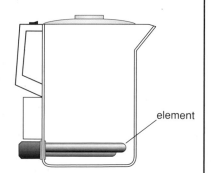

element

3 The freezer compartments of fridges are always found at the top of the fridge. Explain why.

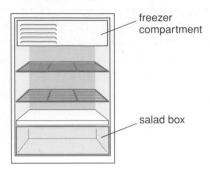
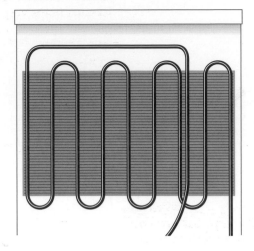

4 As food is cooled in a fridge, heat is transferred from the food to a coolant. This coolant passes through pipes at the back of the fridge. These pipes are painted matt black. Explain why.

5 Large storage containers for gas and liquid fuels are usually painted silver or white. Explain why.

6 Give two reasons, other than they might look good and last a long time, why many people have their windows double-glazed.

7 Cavity walls are often filled with foam to prevent heat transfer through the cavity from a warm room to the air outside. Explain how the foam reduces heat transfer:

a) by conduction
b) by convection.

5.2	**Efficiency**

Co-ordinated	Modular
10.22	Mod 09 10.3

A light bulb in a torch is designed to transfer **electrical energy** to light energy. However, as with most light bulbs, heat energy is also transferred. This heat is not wanted. In most filament bulbs only 5% of the electrical energy is changed to useful light energy. 95% of the electrical energy is transferred as unwanted heat energy. Such light bulbs are described as being only 5% efficient.

Eventually all energy, both useful and unwanted (wasted) is transferred to the surroundings. This makes the surroundings become warmer. A rise in temperature is not usually noticed because the energy becomes spread out. The more spread out it becomes the more difficult it is for any further useful energy transfers.

Efficiency

Efficiency is a way of calculating how good a device is at transferring the total energy going in compared to useful energy transferred. If the total energy going in is the same as the useful energy transferred the device is 100% efficient. No device can have an efficiency greater than 100%.

Efficiency of a device can be calculated using the equation:

$$\text{efficiency} = \frac{\text{useful energy transferred by device}}{\text{total energy supplied to device}}$$

Efficiency of some electrical devices
Low energy light bulbs v filament light bulbs
A 20 W low-energy light bulb gives out as much light as a 100 W filament lamp. In such bulbs very much more of the electrical energy is transferred as light. Low-energy light bulbs are small fluorescent lamps. They do not have a metal filament that gets hot. Very little electrical energy is transferred into unwanted heat, so these light bulbs are more efficient at transferring electrical energy as light energy.

Figure 5.8
a) filament light bulb and b) a low energy light bulb

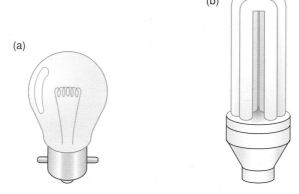

Electric kettles
Electric kettles contain an element that is designed to transfer electrical energy to heat energy. The heat energy makes the water hot enough to boil. The most efficient kettle will be the one in which most heat energy from the element is transferred to the water and not the surroundings.

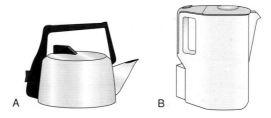

Figure 5.9

Even though each of the kettles shown in Figure 5.9 may have identical elements and hold the same amount of water, they each have some design features that affect the amount of heat lost to the surroundings. These are given in Figure 5.10.

Figure 5.10
Comparing electric kettles

Heat-saving features		Heat-losing features	
Kettle A	**Kettle B**	**Kettle A**	**Kettle B**
Shiny metal surface: this reduces heat loss by radiation	Plastic casing: this reduces heat loss by conduction	Metal casing: this increases heat loss by conduction	Cylindrical shape: this provides a large surface area: volume ratio, so more heat will be transferred to the surroundings through the walls than in the more spherical-shaped kettle

Using energy and doing work

Calculating the efficiency of a kettle

An electric kettle, rated at 2500 W, transfers 2500 J of electrical energy to heat energy each second. The kettle takes 150 s to boil some water. In this time 336 000 joules of energy are transferred to the water. Calculate the efficiency of the electric kettle.

The useful energy output is the heat transferred to the water = 336 000 J

The total energy input is the electrical energy supplied to the element in the kettle. So, total energy input = 2500 × 150

$$= 375\ 000\ \text{J}$$

Substituting in efficiency = $\dfrac{\text{useful energy transferred by device}}{\text{total energy supplied to device}}$

$$= \frac{336\ 000\ \text{J}}{375\ 000\ \text{J}} = 0.9\ \text{J}$$

Multiplying this by 100 produces an efficiency for the kettle of 90%.

So:

- 90% of the electrical energy supplied by the element is used to boil the water.

- 10% of the energy supplied by the element is wasted. Most of this will be transferred as unwanted heat to the surrounding air, the material of the kettle and the element. Some, but not much, will be lost as the boiling water evaporates.

Electric motors

Electric motors are designed to transfer electrical energy to kinetic energy. Inside all electric motors are parts that spin very fast. When these parts spin, friction between the moving parts and the rest of the motor often causes heat energy to be transferred to the motor. The production of heat energy reduces the amount of kinetic energy transferred. This reduces the efficiency of the motor. The more friction, the less efficiency.

Summary

- ◆ The **efficiency** of an energy transfer can be determined by comparing the useful energy output with the total energy input.

- ◆ efficiency = $\dfrac{\text{useful energy transferred by device}}{\text{useful energy supplied to device}}$

Topic questions

1 An electric oven is described as being 70% efficient. what does this mean?

2 A small electric heating element can be used to boil one cupful of water. If the heating element is rated at 300 W, it transfers 300 J of electrical energy to heat energy each second. It takes 6 minutes for the water in the cup to boil and in this time 84 000 J of heat energy are transferred to the water. Calculate the efficiency of this method.

3 Copy and complete the following sentences using words from the box. Each word can be used once or not at all.

chemical	electrical	heat	less	more	wasted

An electric fire is _____ efficient than an electric motor. The fire is designed to transfer _____ energy to heat energy, but the _____ energy produced in a motor is _____ .

162

Energy resources

Producing electricity from renewable and non-renewable energy resources

Some energy resources can be replaced faster than they can be used. These are called **renewable** energy resources. Some energy resources are being used at a faster rate then they are being replaced. These are called **non-renewable** energy resources.

One of the most important uses for renewable and non-renewable energy resources is in the production of electricity. The methods by which some of these resources are used to produce electricity are shown in Figure 5.11.

Figure 5.11

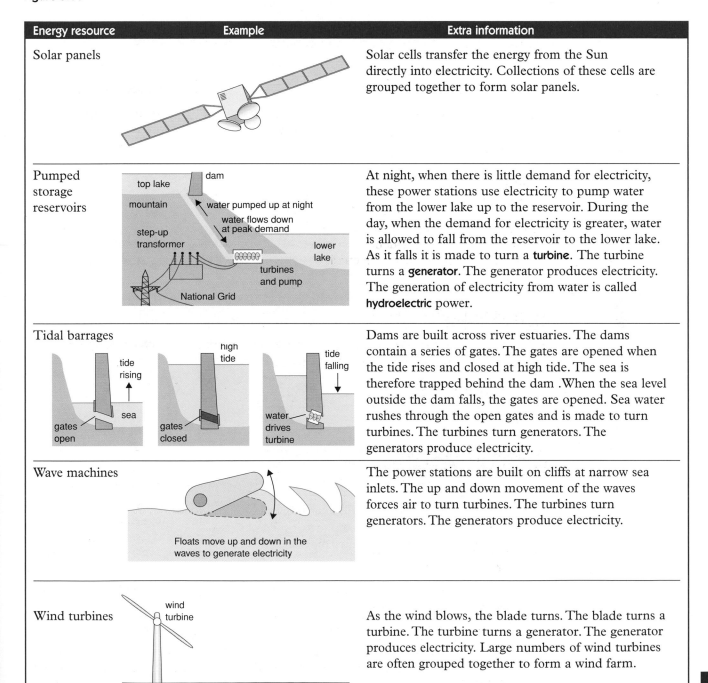

Energy resource	Example	Extra information
Solar panels		Solar cells transfer the energy from the Sun directly into electricity. Collections of these cells are grouped together to form solar panels.
Pumped storage reservoirs		At night, when there is little demand for electricity, these power stations use electricity to pump water from the lower lake up to the reservoir. During the day, when the demand for electricity is greater, water is allowed to fall from the reservoir to the lower lake. As it falls it is made to turn a **turbine**. The turbine turns a **generator**. The generator produces electricity. The generation of electricity from water is called **hydroelectric** power.
Tidal barrages		Dams are built across river estuaries. The dams contain a series of gates. The gates are opened when the tide rises and closed at high tide. The sea is therefore trapped behind the dam .When the sea level outside the dam falls, the gates are opened. Sea water rushes through the open gates and is made to turn turbines. The turbines turn generators. The generators produce electricity.
Wave machines		The power stations are built on cliffs at narrow sea inlets. The up and down movement of the waves forces air to turn turbines. The turbines turn generators. The generators produce electricity.
Wind turbines		As the wind blows, the blade turns. The blade turns a turbine. The turbine turns a generator. The generator produces electricity. Large numbers of wind turbines are often grouped together to form a wind farm.

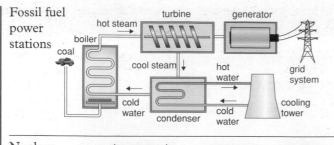

Fossil fuel power stations

The fuel is burned and the heat used to turn water into steam. The steam is made to turn turbines. The turbines turn generators. The generators produce electricity.

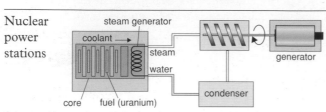

Nuclear power stations

Heat is produced in a nuclear reactor when uranium atoms are made to split. This heat is removed from the reactor by a coolant. The heat in the coolant turns water into steam. The steam is made to turn turbines. The turbines turn generators. the generators produce electricity.

Geothermal energy

The heat energy produced by the natural heating processes in the Earth is called **geothermal energy**. The heat energy is produced in the Earth by the decay of radioactive elements such as uranium. This heat energy is transferred to the lithosphere (see section 3.7) by convection currents in the magma. This heat energy is transferred to any water in the underlying rocks. In areas where there are volcanoes, geysers and hot springs this hot water can reach the surface as steam. This steam can be used directly to the drive turbines that turn the generators that generate electricity.

The advantages and disadvantages of using different energy sources to generate electricity

Fuel	Advantages	Disadvantages	Financial and economic considerations
Fossil fuels – coal, oil and natural gas	Relatively cheap start up and running costs. There are still large reserves in the world although they are getting more expensive to exploit.	When burnt they release carbon dioxide which is a greenhouse gas so will increase the rate of global warming. There is no way to stop the carbon dioxide entering the atmosphere. Burning coal and oil releases sulphur dioxide which causes acid rain. Sulphur can be removed from the fuels before they are burnt or the sulphur dioxide can be removed from the waste gases before they enter the atmosphere. Non-renewable energy resource.	In generating the same amount of heat, coal produces the most carbon dioxide and natural gas the least. Removing the sulphur or the sulphur dioxide increases the cost of the electricity generated. Coal-fired power stations and most oil-fired power stations need to be kept running all the time because the furnaces are damaged if they are allowed to cool down. Many gas-fired power stations can be switched on and off and so are often used when there is a sudden demand for electricity. About 30% efficient.

Fuel	Advantages	Disadvantages	Financial and economic considerations
Nuclear fuels	Do not produce gases that contribute to global warming or the production of acid rain. When working, normally, little or no radiation or radioactive materials are released into the environment.	Radioactive waste is produced and because much of it remains radioactive for a very long time this waste needs to be stored safely and securely, perhaps for thousands of years. In an accident radiation and radioactive materials may be released over a very wide area. Non-renewable energy resource.	Nuclear fuel is relatively cheap, but the cost of building a nuclear power station is very high, as is the cost of closing it down safely (decommissioning) when it is no longer needed. Must be kept running all the time. About 30% efficient.
Wind turbines/ generators	Use a renewable energy resource. Low running costs.	In large groups (wind farms) on hills or cliffs can be unsightly and can cause noise pollution.	The amount of electricity generated is dependent on the strength of the wind and because this varies they are not as reliable as most other sources of electrical energy. These are about 40% efficient at transferring the kinetic energy in the wind into electrical energy.
Tides	Use a renewable energy resource. Low running costs.	Because tidal barrages need to be built across the mouths of rivers they can cause problems for ships and disturb the flow of water. This may destroy the habitats of organisms such as wading birds and the mud-living organisms on which they feed.	Because the amount of electricity produced depends on the tides, which not only vary during each day but from month to month, they are not a reliable source of electrical energy.
Solar cells and solar panels	Use a renewable energy resource. Low running costs. An ideal energy source for producing electricity in remote areas, e.g. on satellites or where only small amounts of electricity are needed, e.g. calculators.	These have the highest start up costs per unit of electricity produced.	The amount of electricity produced depends on the amount of light falling on them. They are not a reliable source of electrical energy. Only about 15% efficient in transferring light energy into electrical energy.
Hydroelectric schemes	Low running costs. Use a renewable energy resource. They can be made to increase their efficiency by being operated in reverse during the night using surplus electricity from other power stations to pump water from the lower reservoir to the higher one.	Many of these schemes involve flooding river valleys – and that was probably not used for farming or forestry but was the habitat for numerous species.	Generally very reliable. Very short start up time, so are often used when there is a sudden demand for electricity. Suitable only for hilly areas with a reliable rainfall.

Summary

◆ The majority of ways in which electricity is generated involve steam/gases/ water turning a **turbine** that turns a **generator**.

◆ The generation of electricity from any energy resource will require choices to be made concerning environmental and economic issues.

Topic questions

1 In what ways is the production of electricity similar in pumped storage power stations, tidal barrages, wave machines and fossil fuel power stations.

2 In what ways is the production of electricity different in a fossil fuel power station and a nuclear power station?

3 Which two gases are produced from a coal or oil-fired power station? What harmful effect does each gas have on the environment?

4 What economic problems need to be considered before building a new nuclear power station?

5 Why does the waste from a nuclear power station need to be stored safely and securely?

6 Why are solar panels a useful energy source for remote areas?

7 How is the efficiency of a hydroelectric power station increased?

8 What environmental problems could be caused by building a new hydroelectric power station?

5.4   Work, power and energy

Co-ordinated	Modular
10.24	Mod 09/11
	10.2,12.2,12.3

Work

Lifting a book from a table requires a small force to be used. To lift the table requires a larger force. In both cases the forces are making the objects move. Forces that cause movement are said to be doing **work**.

Because energy is transferred whenever work is done, energy and work have the same units.

The joule (J) is the unit of work and the unit of energy.

Energy and work are related by the following equation:

$$\text{work done (J)} = \text{energy transferred (J)}$$

This means that if 100 J of work are to be done, then 100 J of energy need to be transferred.

A lot of work will be done and a lot of energy will be required if a very heavy object is to be moved a very long distance.

The amount of work done in moving any object can be calculated using the equation:

$$\text{work done (J)} = \text{force (N)} \times \text{distance moved in the direction of the force (m)}$$
$$W = F \times d$$

Example: A force of 115 N is used to push a loaded wheelbarrow 100 m. Calculate the amount of work done by the force.

work done (J) = force (N) × distance moved in the direction of the force (m)
work done (J) =   115   ×   100
              = 11 500 J

Two people could be doing exactly the same amount of work by pushing identical wheelbarrows the same distance. The person who does the work the quickest is said to be the most powerful.

Power

Power is a measure of how quickly work is done or how quickly energy is being transferred. The unit of power is the watt (W).

The greater the power in a system, the more energy can be transferred in a given time.

Power can be calculated using the equation:

$$\text{power (watt, W)} = \frac{\text{work done (joule, J)}}{\text{time taken (second, s)}}$$

$$P = \frac{F \times d}{t}$$

Example: An electric motor is used to raise a load of 105 N. The motor takes 3 s to lift the load through a vertical distance of 2 m. Calculate the power of the motor.

$$\text{power (W)} = \frac{\text{work done (J)}}{\text{time taken (s)}}$$

$$\text{power (W)} = \frac{105 \times 2}{3}$$

$$\text{power (W)} = \frac{210}{3}$$

$$\text{power} = 70 \text{ W}$$

Example: An electric motor rated at 40 W lifts a load of 80 N through a vertical distance of 1 m in 4 s. Calculate the efficiency of the electric motor.

The equation needed is

$$\text{efficiency} = \frac{\text{useful energy transferred by device}}{\text{total energy supplied to device}}$$

The useful energy output is the energy transferred from the motor to the load.

This can be calculated using the equation:

work done (J) = force (N) × distance moved in the direction of the force (m)
work done (J) =   80   ×   1
              = 80 J

The total energy input is the electrical energy supplied to the motor. This can be calculated by substituting in the equation:

$$\text{power (W)} = \frac{\text{work done (J)}}{\text{time taken (s)}}$$

$$40 = \frac{\text{work done}}{4}$$

So work done (energy transferred) = 40 × 4
= 160 J

To calculate the efficiency, substitute in:

$$\text{efficiency} = \frac{\text{useful energy transferred by device}}{\text{total energy supplied to device}}$$

$$\text{efficiency} = \frac{80}{160} = 0.5$$

Multiplying this by 100 produces an efficiency for the motor of 50%.

Example: An electric kettle is rated at 2000 W. The kettle takes 3.5 minutes to boil some water. In this time 336 000 J of energy are transferred to the water. Calculate the efficiency of the electric kettle.

The equation needed is

$$\text{efficiency} = \frac{\text{useful energy transferred by device}}{\text{total energy supplied to device}}$$

The useful energy output which is the energy transferred from the heating element to the water = 336 000 J.

The total energy input is the electric energy supplied to the kettle.

This can be calculated by substituting in the equation:

$$\text{power (W)} = \frac{\text{work done (J)}}{\text{time taken (s)}}$$

$$2000 = \frac{\text{work done}}{3.5 \times 60}$$

So work done (energy transferred) = 2000 × 210
= 420 000 J

To calculate the efficiency substitute in:

$$\text{efficiency} = \frac{\text{useful energy transferred by device}}{\text{total energy supplied to device}}$$

$$\text{efficiency} = \frac{336\,000}{420\,000} = 0.8$$

Multiplying this by 100 produces an efficiency for the kettle of 80%.

Kinetic energy (KE)

Kinetic energy is the energy possessed by an object due to its motion. A fast moving car will have more kinetic energy than an identical slow moving car. A lorry with a large mass moving at 20 m/s will have more kinetic energy than a car with a small mass moving at 20 m/s.

In order to make the car or the lorry move, energy needs to be transferred from the fuel. The chemical energy in the fuel is transferred to kinetic energy in the moving vehicles.

The kinetic energy of a moving object depends on:

- its mass
- speed

The kinetic energy of a moving object can be calculated using the equation:

$$\text{kinetic energy} = \tfrac{1}{2} \times \text{mass} \times \text{speed}^2$$
$$\text{(joule, J)} \qquad \text{(kilogram, kg)} \quad [\text{(metre/second)}^2, \text{(m/s)}^2]$$

$$\text{k.e.} = \tfrac{1}{2} mv^2$$

Example: A runner of mass 75 kg runs at 10 m/s. What is the kinetic energy of the runner?

Substituting in the equation:

$$\text{kinetic energy} = \tfrac{1}{2} mv^2$$
$$= \tfrac{1}{2} \times 75 \times 10 \times 10$$
$$= 3750 \text{ joules}$$

Gravitational potential energy (PE)

Gravitational potential energy is the energy stored in an object because of the height to which it is lifted against the force of gravity.

A stone thrown 20 m upwards into the air will return to the ground because of the force of Earth's gravitational field on the stone. To throw the stone 20 m into the air requires energy to be transferred from the muscles to the stone. This energy has to overcome the Earth's gravitational pull on the stone.

When the stone is at the top of the throw, all the energy in the stone is called gravitational potential energy (often called potential energy). The larger the stone or the higher it is thrown, the greater the amount of gravitational potential energy that has been transferred to the stone.

When a weight is lifted, work is done against gravity and the gravitational potential energy of the object increases. Because work is being done against gravity it is the vertical height through which the weight is lifted that is measured. So:

$$\text{change in gravitational potential energy} = \text{weight} \times \text{change in vertical height}$$
$$\text{(joule, J)} \qquad \text{(newton, N)} \qquad \text{(metre, m)}$$

But:

$$\text{weight} = \text{mass} \times \text{gravitational field strength}$$
$$\text{(newton, N)} \quad \text{(kilogram, kg)} \quad \text{(newton per kilogram, N/kg)}$$

So,

$$\text{change in gravitational potential energy} = \text{mass} \times \text{gravitational field strength} \times \text{change in vertical height}$$
$$\text{(joule, J)} \quad \text{(kilogram, kg)} \quad \text{(newton per kilogram, N/kg)} \quad \text{(metre, m)}$$

Using energy and doing work

This is often written as:

$$\text{g.p.e.} = mg\Delta h$$

On Earth the gravitational field strength is 10 N/kg.

Figure 5.12

Example: The weightlifter raises a bar of mass 70 kg to a height of 2.5 m.

Calculate the gravitational potential energy of the bar.

$$\text{g.p.e.} = mg\Delta h$$
$$\text{g.p.e.} = 70 \times 10 \times 2.5$$
$$= 1750 \,\text{J}$$

The link between kinetic energy and gravitational potential energy

Figure 5.13 shows the energy transfers that take place when a pendulum swings. At position A, the pendulum is at the end of its swing and it has potential energy. As it moves to position B, its potential energy is transferred to kinetic energy. At C, all its energy has been transferred to kinetic energy.

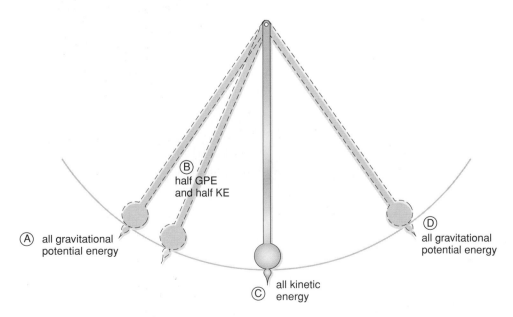

Figure 5.13
Kinetic energy and potential energy for a pendulum

The idea of energy transfer between gravitational potential energy and kinetic energy is used in many calculations.

Example: A stone of mass 1 kg is thrown vertically 30 m into the air. Calculate the gravitational potential energy transferred to the stone when it reaches the top of its path. Ignore air resistance.

$$\text{gravitational potential energy} = mg\Delta h$$
$$\text{gravitational potential energy} = 1 \times 10 \times 30$$
$$= 300\,\text{J}$$

It is also possible to calculate the speed with which the stone was thrown into the air. This is because at the moment it was thrown, its speed was greatest. Therefore all its energy was kinetic energy. As it rose through the air it lost speed as all its kinetic energy was gradually transferred to gravitational potential energy. At the top of its flight, all its kinetic energy had been transferred to gravitational potential energy.

$$\text{So kinetic energy at the moment of being thrown} = \text{gravitational potential energy at the top of the throw}$$

$$\text{So,}\ \tfrac{1}{2}\,mv^2 = mgh$$

Substituting

$$\tfrac{1}{2} \times 1 \times v^2 = 1 \times 10 \times 30$$
$$0.5\,v^2 = 300$$
$$v^2 = 600$$
$$v = 24.5\,\text{m/s}$$

Elastic potential energy

Elastic potential energy is the energy stored in an elastic object when work is done to change its shape. In Figure 5.14 energy from muscles is transferred to the stretched elastic cords.

Figure 5.14
Using a chest expander

Summary

- ◆ Energy is transferred when work is done.

- ◆ The joule (J) is the unit of work and energy.

- ◆ Work done (J) = force (N) × distance moved in the direction of the force (m).

- ◆ Power is the rate of doing work.

- ◆ The watt (W) is the unit of power.

- ◆ Power (W) = $\dfrac{\text{work done (J)}}{\text{time taken (s)}}$

- ◆ **Kinetic energy** is the energy of motion.

  - ◆ Kinetic energy = $\frac{1}{2}mv^2$

- ◆ **Gravitational potential energy** is the energy due to position.

- ◆ Gravitational potential energy = $mg\Delta h$.

- ◆ **Elastic potential energy** is the energy stored when work has been done to change the shape of an object.

Topic questions

1 Write down the equation that links force, distance moved in the direction of the force and work done?

2 a) What are the units of work?
   b) How much work is transferred when a crane lifts a load of 5000 N through a distance of 30 m
   c) When does a force not do any work?

3 a) Write down the equation that links power, time taken and work done?
   b) What are the units of power?

4 What power is being used in each of the following?

   a) A crane lifting a load of 5000 N through a distance of 30 m in 10 s.
   b) An electric motor lifting a load of 50 N over 2 m in 5 s.
   c) A person weighing 550 N running up stairs in 2 s. The stairs are made of 15 steps each 16 cm high.

5 What energy is described as being the energy due to position?

6 A car travelling at a certain speed stops in a distance of 20 m once the brakes have been applied. If the road conditions and braking force remain the same, what will be the stopping distance once the brakes are applied if the car travels at:
   a) twice the original speed?
   b) three times the original speed?
   c) half the original speed?

7 Calculate the kinetic energy for a car of mass 1000 kg travelling at a speed of 25 m/s.

8 Calculate the gravitational potential energy of a mass of 3509 kg that has been lifted by a crane a distance of 20 m. (Take the gravitational field strength = 10 N/kg.)

9 A 0.75 kg ball is thrown vertically into the air. It reaches a height of 15 m. At what speed will it hit the ground? (Take the gravitational field strength as being equal to 10 N/kg and ignore air resistance.)

Examination questions

1 a) Using words or phrases from the list copy and complete the sentences.

   elastic   frictional   gravitational
   less than   more than   the same as

   When a child goes down a slide the _____ force makes him go faster.
   On a damp day the child takes longer to go down the slide. This is because on a damp day the force of friction is _____ on a dry day.
   (2 marks)

   b) Using words or phrases from the list copy and complete the sentence.

   elastic   gravitational potential   sound
   kinetic (movement)   light   thermal (heat)

   When the child goes down the slide, the energy transfers are from _____ energy to _____ energy and _____ energy.
   (3 marks)

2 a) The list gives energy resources which can be used to produce electricity.

   coal   gas   nuclear fuel   oil
   sunlight   tides   waves   wind   wood

   Write down the **four non-renewable** energy resources.

   (4 marks)

   b) Using words from the list copy and complete the sentences about generating electricity.

   energy   gas   generator   smoke
   steam   transformer   turbine   water

   In a coal-fired power station, coal is burnt to release _____. This is used to change _____ into _____ which drives a _____. Electricity is produced by a _____.
   (5 marks)

3 a) A weightlifter has lifted a weight of 2250 newtons above his head. The weight is held still.
   i)   In the box are the names of three forms of energy.

gravitational potential   kinetic   sound

   Which one of these forms of energy does the weight have?   *(1 mark)*
   ii) What force is used by the weightlifter to hold the weight still?
        Give a reason for your answer.   *(2 marks)*

   b) To lift the weight, the weightlifter does 4500 joules of work in 3.0 seconds.
   Use the following equation to calculate the power developed by the weightlifter. Show clearly how you work out your answer.

   $$\text{power} = \frac{\text{work done}}{\text{time taken}}$$

   (2 marks)

4 The diagram shows a high jumper.

In order to jump over the bar, the high jumper must raise his mass by 1.25m.
The high jumper has a mass of 65kg. The gravitational field strength is 10 N/kg.
a) The high jumper just clears the bar.
   Calculate his gravitational potential energy.
   (4 marks)
b) Calculate the minimum speed the high jumper must reach for take-off in order to jump over the bar.   *(3 marks)*

5 The drawing shows an investigation using a model steam engine to lift a load.

In part of the investigation, a metal block with a weight of 4.5 N was lifted from the floor to a height of 90 cm.

a) i) Calculate the work done in lifting this load. Write the equation you are going to use, show clearly how you get to your answer and give the unit. *(3 marks)*

ii) How much useful energy is transferred to do the work in part a) i)? *(1 mark)*

b) In another part of the investigation, 250 J of work is done in one minute.

Calculate the useful power output. Give the unit. *(2 marks)*

6 State and explain the advantages and disadvantages of using nuclear power stations to produce electricity compared with coal-fired power stations. *(4 marks)*

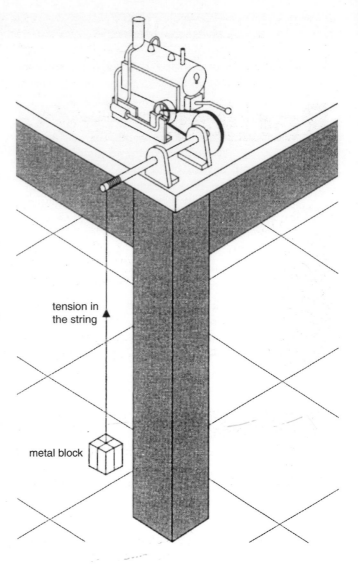

tension in the string

metal block

Chapter 6
Radioactivity

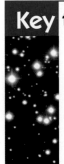
6.1		Types, properties and uses of radioactivity
Co-ordinated	**Modular**	
10.27	Mod 12 13.3	

Radioactivity is the name given to the particles and rays that come from the unstable **nuclei** of certain elements. Radioactivity cannot be seen, heard or felt but it will affect photographic film. This is how it was first discovered in 1896. The rays and particles are emitted at **random** from a radioactive substance. Heating the substance or trying to dissolve it or combining it chemically with another will not affect its radioactivity.

As an unstable nucleus emits a particle or ray, it changes. This process is called **radioactive decay**.

Figure 6.1
Unstable nucleus emitting a particle and a ray

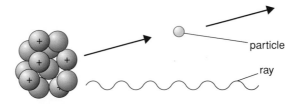

Detecting radioactivity

The average number of emissions in a certain time is called the **activity** or **count rate**. The activity depends only on the number of **atoms** of the particular radioactive element present. As the unstable nuclei **decay**, the activity decreases because fewer atoms of the original radioactive element are left.

Figure 6.2
As unstable atoms break down, the level of radiation falls

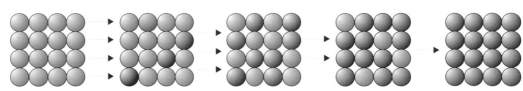

 unstable (radioactive atom)    stable (decay product)   ▶ radiation

Radioactivity

The activity can be measured by a **Geiger–Müller tube** and counter.

Figure 6.3
A Geiger–Müller tube counting radiation

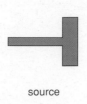
source

absorber

Geiger–Müller tube

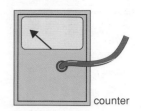
counter

Figure 6.4
A radiation-detecting badge

People who work with radioactive substances wear a special badge containing photographic film. The badge is processed regularly to measure the amount of **radiation** received. This is done to check that the badge wearer has not been exposed to too much radioactivity. The more the exposure to radiation, the darker the film appears after it has been developed.

Did you know?

Henri Becquerel discovered the radioactivity of uranium in 1896. He was awarded the Nobel Prize for Physics in 1903 jointly with Pierre and Marie Curie. The Curies isolated the radioisotopes radium and polonium from uranium ore. They worked on a small budget in dilapidated surroundings. Marie Sklodowska Curie continued this work after her husband's death. She received the Nobel Prize for Chemistry in 1911.

Henri Becquerel

Pierre and Marie Curie

Sources of radiation

Radioactive emissions are classed as either natural or man-made. There is a lot of radioactivity from natural sources such as the rocks in the Earth's crust and from deeper inside the Earth. One of the radioactive elements released from rocks is radon gas. This can accumulate to dangerous levels in confined spaces. Certain surface rocks, such as the Cornish and Aberdeen granites, show more activity than others.

Cosmic rays, which constantly bombard the Earth from space, are a major source of the radioactivity in the air.

Man-made sources are the results of the processing and use of radioactive materials in industry, medicine, nuclear reactors and some weapons. Highly radioactive materials are strictly contained and their disposal is controlled.

Radioactive waste material from nuclear reactors, factories and hospitals is either specially stored or, if it has little radioactivity, it is diluted and discharged into the environment. These man-made sources make only a small contribution to the natural sources of radioactivity.

Figure 6.5
Different sources of radiation: a) medical (mainly X-rays), b) air flights increase exposure to cosmic rays

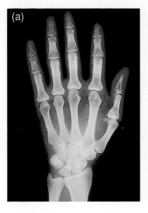

So the air we breathe, the food we eat, the rocks around us, even our bones, become sources of radioactivity. These radiations which are around us all the time are called **background radiation**. The level of background radiation changes from place to place but is always quite low. Occasionally people receive more radiation as a result of their work or from medical treatments. Nuclear accidents are rare but do raise radiation levels at those sites.

Figure 6.6 ▲
The storage of radioactive material

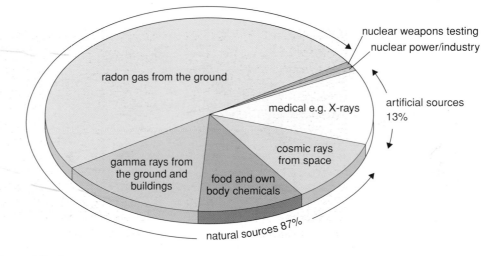

Figure 6.7 ▲
Typical radiations for the UK (NRPB)

Did you know?

The Chernobyl nuclear reactor became supercritical when its reaction went out of control in April 1986. Eventually it caught fire. The remnants are still very radioactive and will be for a long time yet. The shield of thick concrete extends under the reactor. Despite the hazards, courageous engineers continue to monitor the levels of radioactivity and check the safety of this concrete 'tomb'.

Types of emission

When radioactivity was first discovered, the identity of the different particles and rays was not known. They were called **alpha** (α), **beta** (β), and **gamma** (γ). (These are the first three letters of the Greek alphabet.) What they are and how they behave is now better understood.

Alpha particles are quickly stopped by thin paper or a few centimetres of air. Alpha particles easily **ionise** atoms by knocking out the outer **electrons** from the atoms of the absorbing material. With each collision the alpha particle loses energy and so it does not travel far. It will not travel more than 5 cm in air.

Beta particles are very much smaller than alpha particles. They travel further at high speeds through other substances before losing their energy by colliding with the atoms of that substance. They ionise matter less well than alpha particles as they are so small. Beta particles can be stopped by a 5 mm thick sheet of aluminium.

Gamma rays are not charged particles like alpha and beta particles, but are high-energy waves belonging to the electromagnetic spectrum. They are very penetrating as they are poor ionisers and do not lose their energy quickly. Gamma radiation is never entirely stopped, but can be reduced to almost zero levels by a 10 cm thick sheet of lead or metres of concrete.

Figure 6.8
Selective absorption of radioactive emissions

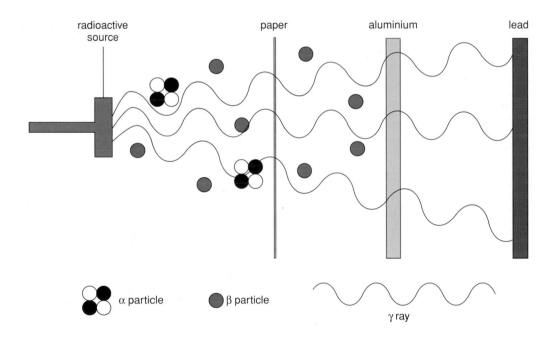

When radioactive substances need to be shielded they are placed in thick lead cans or stored behind several metres of concrete. This blocks almost all the radiation.

Gamma rays have very short wavelengths, shorter than those of X-rays. Both types of wave belong to the electromagnetic spectrum, both are penetrating and poor ionisers, but they are produced in different ways. X-rays are not radioactive emissions, as they do not come from the nucleus of the atom.

Figure 6.9
Radioactive material is transported in this thick metal container

Identifying emissions

The type of emission from a source can be found by measuring the radioactivity level with various absorbers placed between the source and the detector.

The results in Figure 6.10 show that paper does not reduce the count rate very much, so there are no alpha particles emitted. Aluminium reduces the count rate considerably – this suggests that beta particles are emitted. Lead reduces the count rate down to almost the level of background radiation (0.4 counts/second) so gamma rays are emitted.

Figure 6.10
Identifying radiation from an unknown source, X

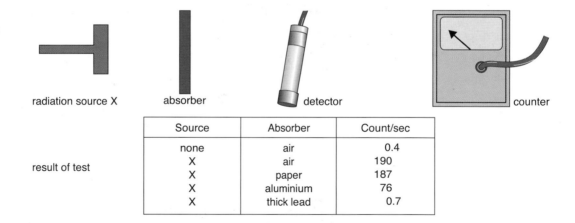

radiation source X   absorber   detector   counter

result of test

Source	Absorber	Count/sec
none	air	0.4
X	air	190
X	paper	187
X	aluminium	76
X	thick lead	0.7

Radiation and living cells

Particles and rays come out of the nucleus with great energy allowing them to pass through other substances. As the radiation is absorbed, it ionises molecules of that substance, causing the substance to change. This is why radioactivity can be dangerous to living cells, because their chemical structure, particularly their DNA, is altered and those cells do not work properly any longer. In certain cases, the cells are so changed that they become a threat to the living organism, affecting the normal cells and causing cancers (see section 3.3).

Radioactive sources are handled according to strict government guidelines. Great care is taken to prevent penetrating emissions, such as beta particles or gamma rays, from reaching the body from an external source. Protective clothing and special gloves stop beta and gamma radiation travelling through the skin and reaching the internal organs.

Radioactivity

Alpha particles are unlikely to penetrate the skin. However if a radioactive source is taken internally, then alpha particles are most dangerous as they are powerful ionisers of matter.

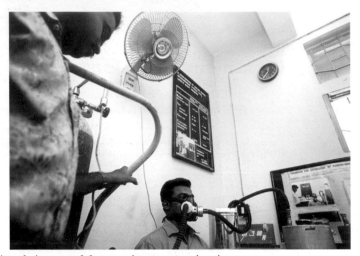
Uses of radiation

Harmful bacteria in food can be killed by exposing the fresh food to gamma radiation. This prevents the bacteria from multiplying and spoiling the food. Germs on hospital instruments and dressings, which could infect patients, are made harmless by gamma radiation. This method of sterilising surgical instruments is useful when boiling in water could damage the instrument.

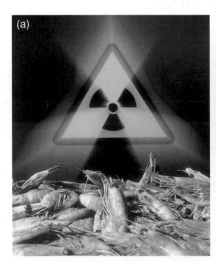

Figure 6.11
Some uses of radiation:
a) harmful bacteria in food can be killed by gamma rays,
b) operating equipment is sterilised by gamma rays

Radiation can be used to cure cancer. It may seem strange that radioactivity can cause cancer and cure it. In radiotherapy large doses of radioactivity are carefully given from an external source of gamma rays so the cancerous area receives the most radiation and nearby healthy tissue is less affected (see section 3.3).

Figure 6.12
Radiotherapy apparatus

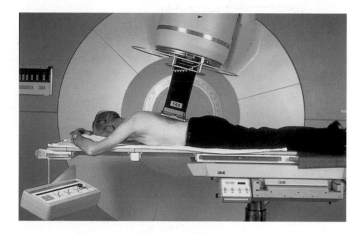

For some conditions the radioactive isotope is given internally. For this method to work, the radioactive isotope must go to the particular part to be treated. The substance used is usually a beta emitter so that the effects are localised. Gamma rays would penetrate too far and could affect healthy tissue.

Figure 6.13
Iodine-131 localised in the thyroid gland

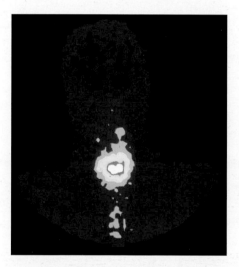

Radioactive sources are commonly used in medical tests. Injecting a person with iodine-131 can show if that person has an enlarged thyroid gland. Radioactive isotopes with fast decay rates are chosen for internal doses. This is to prevent long-term radiation affecting normal, living cells.

Industrial uses of radiation

If a **radioactive tracer** is put in a fluid, the path of the fluid can be followed even when it cannot be seen. Gamma rays from the tracer will penetrate pipes and soil to reach a detector. Such tracers are used to test liquid flow rates through pipes, the dispersion of effluents into larger quantities of water, the uptake of fertilisers by plants and leaks in underground pipelines (Figure 6.14).

The tracers are chosen with care not to damage the environment. The element is usually non-toxic, it is well diluted and its radioactivity decays quickly.

Radioactive isotopes can also be used industrially to control the quality of material. For example the thickness of an aluminium sheet can be controlled by measuring how much beta radiation passes through it. If the amount of radiation is too high, the sheet is too thin. The position of the rollers can then be automatically adjusted to produce the correct thickness (Figure 6.15).

Figure 6.14
Using radioactive tracers to locate a leak in a pipe

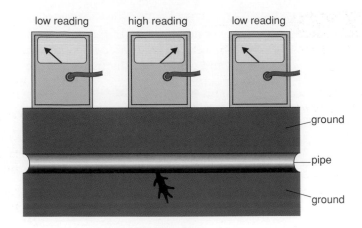

Figure 6.15
Using a beta source to control the thickness of an aluminium sheet

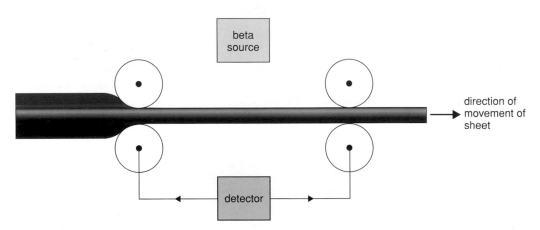

Summary

◆ **Radioactivity** is the random emission of particles and rays from an unstable **nucleus**. It can be detected by photographic film or a **Geiger–Müller tube**.

◆ The radiation from space, food , rocks, air and buildings is called **background radiation**.

◆ Radiation can damage living cells.

◆ Radiation has a number of medical and industrial uses.

Type of emission	Nature	Penetrating ability
alpha	large, + charged particle	5 cm in air; stopped by paper
beta	small, − charged particle	50 cm in air; stopped by 5 mm thickness of aluminium foil
gamma	high frequency electromagnetic wave	intensity falls with distance; blocked by 10 cm thickness of lead

Topic questions

1 Copy and complete the sentences, using some of these words

> temperature stable different chemical
> unstable particle random atom nucleus

Radiations come from an _____ nucleus. The activity is _____ and cannot be affected by _____ , pressure or _____ reactions. When a _____ leaves a nucleus, the _____ changes to that of a _____ atom.

2 Can you find the radioactive elements from these nonsense words?

MONOPULI DIAMUR RUINAUM
DROAN RIMOUTH

3 Copy and complete these sentences.
   a) The activity of all radioactive materials _____ with time.
   b) _____ particles travel a short distance in air and are easily _____ .

c) _____ rays travel from space and are a _____ source of radiation.
d) Radioactive _____ are used in medicine to treat _____ .

4 Put the following sources of radiation into a table, under the correct heading, Natural or Man-made:

building materials, cosmic rays, X-rays, plants, radon gas, nuclear waste.

5 A source was examined for the type of emissions it produced. Find the emissions from the results in table below.

no absorber	360 counts/sec
paper absorber	180 counts/sec
aluminium absorber	85 counts/sec
lead absorber	3 counts/sec

6.2		**Atomic structure and radioactivity**

Co-ordinated	Modular
10.28	Mod 12 13.4

Radioactivity is the result of changes in the nuclei of atoms. Only after the structure of the atom had been discovered could the various forms of radiation be explained.

The story of the atom – John Dalton to Ernest Rutherford

In the early part of the 19th century John Dalton had proposed that the smallest part of an element was a tiny solid indivisible particle called an atom. In 1897, J J Thomson discovered the existence of a particle much smaller than an atom. This particle was called an electron. Thomson knew that because electrons have a negative charge the rest of the atom should have a positive charge if the atom was to be held together. In 1904 he proposed what has been called the 'plum-pudding' model of the atom. It was given this name because the model looked like a pudding, which represented a sphere of positively charged material, with bits of plum, representing the negatively charged electrons, scattered in the pudding.

In 1906, Ernest Rutherford discovered the existence of alpha particles. These particles had a double positive charge and so were thought to be the positive parts of the Thomson atom.

Ernest Rutherford investigated his idea for the structure of the atom with Hans Geiger and Ernest Marsden, at Manchester University between 1909–1911. They fired a thin beam of alpha particles, which they knew had a double postive charge, at very fine gold foil (Figure 6.17). Instead of passing straight through the foil, as would be expected if the atom was like the one proposed by Thomson, the alpha particles were scattered like the light from a torch.

blob of positive charge

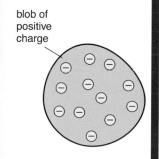

Figure 6.16
J J Thomson's 1904 'plum pudding' model of the atom

Figure 6.17
*How they set up
their experiment*

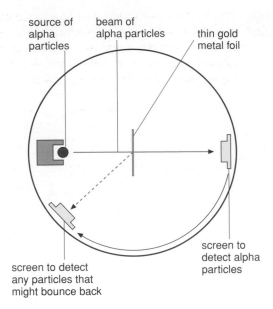

The results, taken over many months, when more than 100 000 measurements were made, showed that whilst most alpha particles went straight through the foil, a few were deflected through quite wide angles and some even bounced back towards the source. These unexpected results came as a surprise.

Figure 6.18
*What the results
showed*

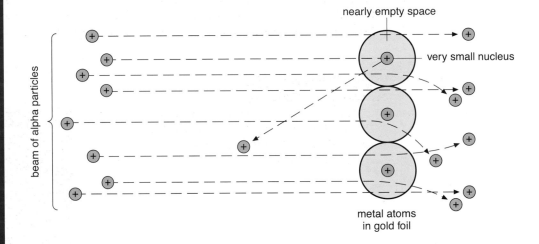

Rutherford decided that:

- The alpha particles which were not deflected had been unaffected by the atoms in the gold foil and were going through empty space. Because of this the atoms in the foil must contain a lot of empty space.

- The positively charged alpha particles that were deflected had been repelled by the positively charged matter within each gold atom. Because of this the parts of the atom containing the positive charges must be concentrated in a very small, dense part of the atom – the nucleus.

- Any alpha particle that had bounced back must have collided with the nucleus of one of the gold atoms. Because few alpha particles bounced back the nucleus must be very small.

- The electrons orbited around in the empty space surrounding the nucleus.

Rutherford concluded that atoms are made up of a dense positively charged nucleus around which was empty space in which orbited the negatively charged electrons. He also concluded that the amount of positive charge in the nucleus should equal the negative charge on the electrons. Rutherford's atom was the same size as Thomson's but contained mainly empty space.

Figure 6.19
The Rutherford model of the atom

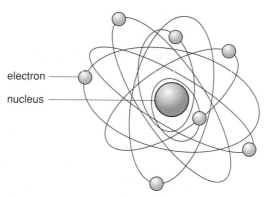

electron

nucleus

The Rutherford Model of an atom
A miniature solar system with the electrons moving like planets around the nucleus.

Rutherford presented his findings on the structure of the atom in 1911. The clear objectives and thorough methods of his experiments convinced many scientists that his idea for a nuclear atom was correct. Niels Bohr incorporated Rutherford's conclusions into a theoretical framework that explained how the electrons were able to orbit a central positive nucleus. Bohr used the light spectrum of hydrogen, the simplest atom, to confirm that electrons were able to orbit the nucleus without falling into it. The nuclear atom became accepted as the correct structure for the atom, and it is still the view we have today, with some modified details.

Did you know?

Joseph John Thomson, a British physicist (1856–1940) found the charge to mass ratio for the electron. He won the Nobel prize for physics in 1906.

Joseph John Thomson

Ernest Rutherford (1871–1937) was born in New Zealand of a Scottish father. He was considered the greatest experimental physicist of his generation and won the Nobel prize for chemistry in 1908.

Ernest Rutherford

Hans Geiger (1882–1945), a German nuclear physicist who took part in many of Rutherford's experiments, also developed a better instrument for detecting radiation, called the Geiger–Müller tube in 1928.

Hans Geiger

Ernest Marsden (1889–1970) a British physicist.

Ernest Marsden

Niels Bohr (1885–1962), a Danish theoretical physicist, the 'father of atomic theory'. He won the Nobel prize for physics in 1922.

Niels Bohr

Radioactivity and the atom

Atomic structure

The nucleus of an atom contains **nucleons**, which are closely packed **protons** and **neutrons**. The protons have a positive charge and the neutrons have no charge. The electrons orbiting the nucleus each have a negative charge of the same size as that on a proton. The number of protons and electrons is the same, making a single atom electrically neutral. Atoms of the same elements which have different numbers of neutrons but the same number of protons are called **isotopes**.

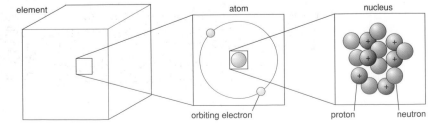

Figure 6.20
Atomic structure

The mass of a neutron is almost the same as that of a proton, and these masses are 2000 greater than that of an electron. As these are such tiny particles, it is easier to talk of their mass in units, which compare the mass of each sub-particle, to that of the simplest atom hydrogen.

On this scale:

sub-atomic particle	mass	charge
proton	1	+1
neutron	1	0
electron	negligible	−1

Because of the work on atomic structure it is now known that radioactivity occurs because of changes in the unstable nuclei of atoms. Atoms with unstable nuclei are called radioactive isotopes (or **radioisotopes** or **radionuclides**). Radiation is produced when an unstable nucleus emits a particle or ray (or both) in order to become more stable. Once a particle has been emitted by the nucleus, then the nucleus changes to that of another element. The number of protons in the new nucleus is different.

The forces which hold the nucleus together are very strong, so any particles that come from a nucleus have large amounts of energy – rather like a bullet from a gun.

Figure 6.21
Unstable nucleus emitting a particle and a ray

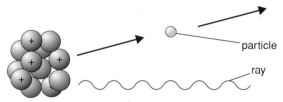

The rays and particles are emitted totally at random from a radioactive substance.

Types of emission

Alpha and beta particles

Alpha particles are the largest particles to be emitted from an unstable nucleus. Alpha particles consist of two protons and two neutrons. This is the same structure as the nucleus of helium.

Figure 6.22
The structure of alpha and beta particles

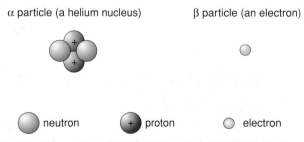

Each alpha particle has a **mass number** of 4 and an **atomic number** of 2. The atomic number is the number of protons, and the mass number, or nucleon number, is the number of protons plus the number of neutrons. When an alpha particle is emitted from an unstable nucleus, the atomic number of the original nucleus goes down by 2. This produces the nucleus of the element two places lower in the periodic table. The mass number decreases by 4. For example, the nucleus of a radium atom decays into radon and an alpha particle. This is represented by the following equation:

$$^{226}_{88}\text{Ra} \rightarrow\ ^{222}_{86}\text{Rn}\ +\ ^{4}_{2}\text{H}$$

Beta particles are fast moving electrons. As an electron is emitted from an unstable nucleus, a neutron changes to a proton. The emission of a beta particle produces a nucleus of an element one place higher in the periodic table. For example, the nucleus of a radioactive carbon atom decays into nitrogen and an electron. This is represented by the following equation:

$$^{14}_{6}C \rightarrow \ ^{14}_{7}N + \ ^{0}_{-1}e$$

Gamma radiation

Gamma radiation is short wavelength electromagnetic radiation (see section 3.3). It is emitted when an unstable nucleus loses excess energy. The emission of gamma radiation causes no change to either the mass number or the atomic number of the nucleus.

Summary

◆ Radiation occurs because of changes in unstable nuclei.

◆ The work of Rutherford helped to develop the modern ideas about the structure of the atom.

◆ Alpha particles are helium nuclei – two protons and two neutrons.

◆ Beta particles are electrons that come from the nucleus. As an electron is released, a neutron becomes a proton.

◆ Gamma radiation is high energy, short wavelength electromagnetic waves.

Topic questions

1 Match the conclusion with the observed result for Rutherford's experiments on alpha scattering.

Conclusion	Observations
Atoms mainly empty space	Curved paths of deflected particles
Concentrated positive nucleus	Most particles passed straight through
Electrostatic repulsion laws obeyed	Some particles bounced back with no energy loss

2 What are the differences between the atomic models of Thomson and Rutherford?

3 Copy and complete the gaps in the following table about sub atomic particles.

sub-atomic particle	mass	charge
	1	+1
neutron		
	negligible	

4 What is a radioisotope?

5 a) An isotope of radium is written as $^{226}_{88}$Ra. Find the proton and neutron numbers.

   b) An isotope of calcium (Ca) has 20 protons and 24 neutrons. Write this in symbol form.

6 Copy and complete the gaps in the table about radioactive emissions.

type of emission	what it is	charge	absorbed by	causes ionisation
	two neutrons and two protons		_____ of air	very strong
beta particle			_____ of aluminium	
gamma ray		no charge	_____ of lead	

6.3 Half-life

The **half-life** of a radioisotope (radionuclide) is the time taken for half the unstable atoms to decay. That is the time it takes for the activity to fall to half its original value. It doesn't matter how much of the radioisotope there is to start with, the half-life is the same. Isotopes with a short half-life give off their radiation more quickly. Therefore they are more dangerous. Figure 6.23 below gives some data for the radioactive decay of an isotope with a half-life of two hours.

Figure 6.23

	Activity in counts/min	Time in hours		
	1800	0		
	1300	1		
(1800/2)	900	2	one half-life	2 hours
	650	3		
(900/2)	450	4	two half-lives	2 × 2 = 4 hrs
	320	5		
(450/2)	225	6	three half-lives	3 × 2 = 6 hrs
	160	7		

This data is shown as a graph in Figure 6.24.

Figure 6.24
The radioactive decay curve for a substance with a half-life of 2 hours

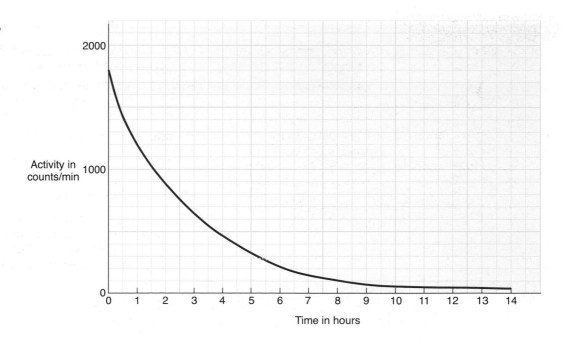

Sources which are very radioactive have a fast decay rate, which means they have a short half-life. Less active sources have a slow decay rate, and a long half-life.

Radioactive dating

The radioactive isotope, carbon-14, is a tiny part of all the carbon taken in by living organisms, which is mostly carbon-12. The proportion of carbon-14 to carbon-12 is constant during the lifetime of an organism. When the organism dies, this fraction decreases as the carbon-14 decays. By comparing the activity for the carbon of a dead sample, to that for living things, the age of the sample can be found. The half-life for carbon-14 is 5730 years. Items over 10 000 years old are not reliably dated by their activity as it is too small to detect and recent items have an activity rate similar to that for living things.

Another method measures the proportion of carbon-14 atoms to carbon-12 atoms directly. This method destroys less of the original sample and can date items up to 100 000 year of age, but it needs special equipment.

Figure 6.25
The age of Lindow Man was found by radiocarbon dating

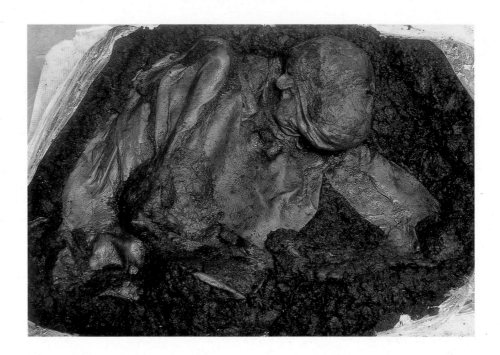

Rocks can be dated by comparing the proportion of uranium and lead atoms in a sample. The half-life of uranium-238 is 4500 million years so this radioisotope can be used to date some of the Earth's oldest rocks.

The radioisotope potassium-40 decays to produce argon as the stable product. The proportion of potassium-40 and argon can be used to date igneous rocks in which the argon has been trapped.

Summary

◆ Very active sources decay very quickly.

◆ The **half-life** of a **radioisotope** is the time taken for half the unstable nuclei to decay to nuclei of different atoms, or for its activity to fall to half its original value. A short half-life indicates a very active source.

◆ Radioactive decay can be used to date materials.

Topic questions

1 A smoke detector uses a radioisotope to ionise the air inside it. This allows a small electric current to flow. In the presence of smoke, the current falls and sets off the alarm.
   a) Which is the best choice of radioisotope given in the table below?
   b) Explain your choice.

Radioisotope	Emission	Decay rate
cobalt-60	gamma	fast
americium	alpha	slow
iodine-131	beta	very fast
technetium-99	gamma	very fast

Activity in counts/min	Time in days
8000	0
6100	10
4600	20
3500	30
2700	40
2000	50
1500	60
1200	70
870	80

2 The table shows the rate of decay for a radioisotope. What is the half-life of this radioisotope? You may plot a graph.

3 The half life of carbon-14 is 5700 years. The activity of the carbon in a living sample is 15 counts/minute. An ancient axe-handle sample of the same mass gives an activity of 3.75 counts/minute. What age is this handle?

4 A radioisotope of lead has a half-life of 10.6 hours. A small sample of lead containing the isotope has a count of 6000 counts per minute. How long will it be before the count rate reaches 375 counts per minute?

6.4   Nuclear fission

Co-ordinated	Modular
10.28	Mod 12 13.4

Producing the heat energy in a nuclear reactor

Atoms, such as those of uranium-235, are unstable. If a slow moving neutron is captured by an atom of uranium-235 then the large nucleus breaks into two smaller parts and some neutrons. This splitting up releases a huge amount of energy and is called **nuclear fission**. This energy is very many times greater than the energy released in a chemical reaction when bonds between two atoms are made. The fission of 1 kg of uranium-235 releases more energy than the burning of 2 000 000 kg of coal. It is the energy released from such a chain reaction that provides the heat energy in the core of a nuclear power station.

Figure 6.26

A neutron hitting the nucleus of an atom of uranium-235

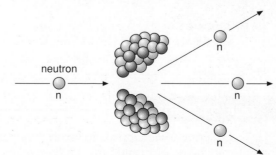

Radioactivity

The released neutrons may then go on to to be captured by more uranium atoms. This makes them split and fire off more neutrons which may be captured by other atoms and so on. This is called a **chain reaction**. New atoms formed by nuclear fission are radioactive.

Figure 6.27
The chain reaction

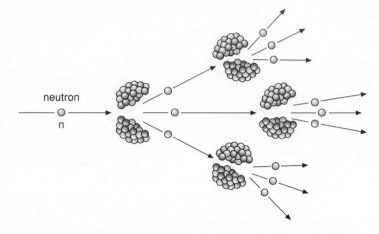

The products of nuclear fission are themselves highly radioactive. When the uranium atom splits, it usually does so in two unequal parts, with a few neutrons. The two new atoms carry more neutrons than their stable isotopes, so they emit β particles rapidly. The fission products are shielded and cooled whilst the more active elements decay. Then they are stored in protective containers before being reprocessed.

Summary

◆ In a nuclear power station the heat is produced by **nuclear fission**.

◆ During nuclear fission a vast amount of energy is released together with two smaller nuclei and several neutrons. The neutrons can split nuclei of other large atoms creating a **chain reaction**.

Topic questions

1 Describe what happens to an atom of uranium-235 when it captures a slow moving neutron and how this leads to a chain reaction.

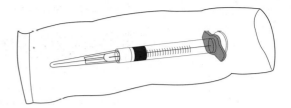

Examination questions

1 a) The different sources of radiation to which we are exposed are shown in the pie chart. Some sources are natural and some artificial.

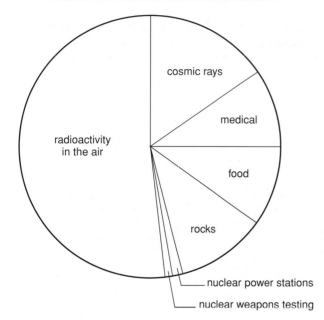

i) Name *one* natural source of radiation shown in the pie chart. *(1 mark)*

ii) Name *one* artificial source of radiation shown in the pie chart. *(1 mark)*

b) A radioactive source can give out three types of emission: alpha particles, beta particles, gamma radiation.
The diagram shows the paths taken by the radiation emitted by two sources, X and Y. What types of radiation are emitted by each of the sources? *(2 marks)*

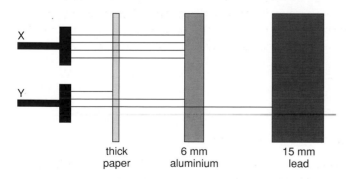

c) The diagram shows a disposable syringe sealed inside a plastic bag. After the bag has been sealed the syringe is sterilised using radiation. Explain why radiation can be used to sterilise the syringe. *(3 marks)*

2 a) The diagram shows the apparatus used by a teacher to investigate an alpha (α) source.

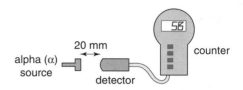

i) Which piece of apparatus could be used as a radiation dectector?

| Geiger-Müller Tube | Oscilloscope | Voltmeter |

(1 mark)

ii) Copy and complete the following sentence.
When a piece of paper is placed between the detector and the alpha source the count rate will go _____ .
(1 mark)

b) Two sheets of steel were joined together by welding.

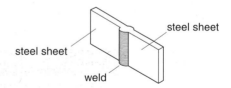

Radiation was used to check how well the welding had been done.

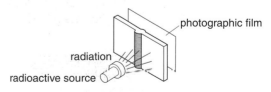

i) Which type of radiation should be used? Give a reason for your answer. *(2 marks)*

ii) The diagram shows the exposed photographic film.

Does the photographic film show that the weld was good or bad? Give a reason for your answer. *(2 marks)*

3 The diagram shows a film badge worn by people who work with radioactive materials. The badge has been opened. The badge is used to measure the amount of radiation to which the workers have been exposed.

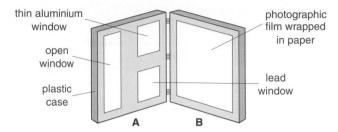

thin aluminium window

open window

plastic case

photographic film wrapped in paper

lead window

A **B**

The detector is a piece of photographic film wrapped in paper inside part **B** of the badge. Part **A** has "windows" as shown.

a) Use words from the list to complete the sentences.

alpha   **beta**   **gamma**

When the badge is closed
i) _____ radiation and _____ radiation can pass through the open window and affect the film. *(1 mark)*
ii) _____ radiation and _____ radiation will pass through the thin aluminium window and affect the film. *(1 mark)*
iii) Most of the _____ radiation will pass through the lead window and affect the film. *(1 mark)*

b) Other detectors of radiation use a gas which is ionised by the radiation.
i) Explain what is meant by *ionised*. *(1 mark)*
ii) Explain why ionising radiation is dangerous to people who work with radioactive materials. *(2 marks)*

4 a) The table gives information about five radioactive isotopes.

Isotope	Type of radiation emitted	Half-life
Californium-241	alpha (α)	4 minutes
Cobalt-60	gamma (γ)	5 years
Hydrogen-e	beta (β)	12 years
Strontium-90	beta (β)	28 years
Technetium-99	gamma (γ)	6 hours

i) What is an alpha (α) particle? *(1 mark)*
ii) What is meant by the term half-life? *(1 mark)*
iii) Which **one** of the isotopes could be used as a tracer in medicine? Explain the reason for your choice. *(3 marks)*

b) The increased use of radioactive isotopes is leading to an increase in the amount of radioactive waste. One method for storing the waste is to seal it in containers which are then placed deep underground.

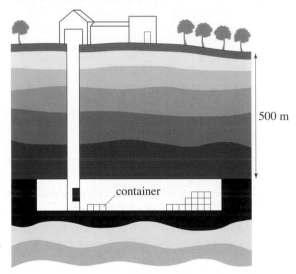

500 m

container

Some people may be worried about having such a storage site close to the area in which they live. Explain why. *(3 marks)*

5 a) The graph shows how a sample of barium-143, a radioactive *isotope* with a short *half-life*, decays with time.
i) What is meant by the term *isotope*? *(1 mark)*
ii) What is meant by the term *half-life*? *(1 mark)*
iii) Use the graph to find the half-life of barium-143. *(1 mark)*

b) Humans take in the radioactive isotope carbon-14 from their food. After their death, the proportion of carbon-14 in their bones can be used to tell how long it is since they died. Carbon-14 has a half-life of 5700 years.
i) A bone in a living human contains 80 units of carbon-14. An identical bone taken from a skeleton found in an ancient burial ground contains 5 units of carbon-14. Calculate the age of the skeleton. Show clearly how you work out your answer. *(2 marks)*
ii) Why is carbon-14 unsuitable for dating a skeleton believed to be about 150 years old? *(1 mark)*

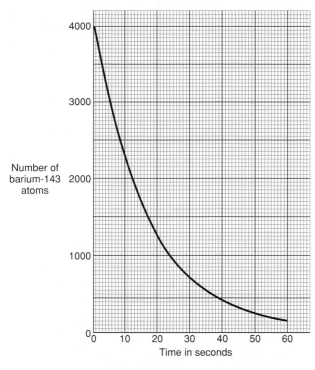

Number of barium-143 atoms vs Time in seconds

a) Draw a graph of the results and find the half-life for the isotope. On the graph show how you obtain the half-life. *(4 marks)*

b) Sodium-24 decays by beta emission. The G-M tube used in the experiment is shown in the diagram. Each beta particle which gets through the glass causes a tiny electric current to pass in the circuit connected to the counter.

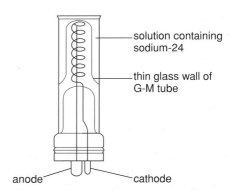

solution containing sodium-24

thin glass wall of G-M tube

anode   cathode

i) Why must the glass wall of the G-M tube be very thin?

ii) Why is this type of arrangement of no use if the radioactive decay is by alpha emission? *(1 mark)*

c) The increased industrial use of radioactive materials is leading to increased amounts of radioactive waste. Some people suggest that radioactive liquid waste can be mixed with water and then safely dumped at sea. Do you agree with this suggestion? Explain the reason for your answer. *(3 marks)*

6 The isotope of sodium with a mass number of 24 is radioactive. The following data were obtained in an experiment to find the half-life of sodium-24.

Time in hours	Count rate in counts per minute
0	1600
10	1000
20	600
30	400
40	300
50	150
60	100

c) Sodium chloride solution is known as saline. It is the liquid used in 'drips' for seriously-ill patients. Radioactive sodium chloride, containing the isotope sodium-24, can be used as a tracer to follow the movement of sodium ions through living organisms. Give one advantage of using a sodium isotope with a half-life of a few hours compared to using an isotope with a half-life of:

i) five years *(1 mark)*

ii) five seconds. *(1 mark)*

Glossary

Acceleration Rate of change of velocity. Units are m/s^2.

Activity/count rate The number of radioactive emissions in a certain time.

Air resistance (drag) The force from the air that opposes movement.

Alpha (radiation/particles) A type of radioactive emission with low penetrating power blocked by paper.

An alpha particle is made up of two neutrons and two protons (a helium nucleus).

Alternating current (a.c.) A current that changes direction as the supply voltage changes from + to −, or − to +.

Ammeter An instrument used to measure the size of an electric current.

Ampere (amp) The unit of electric current.

Amplitude The maximum displacement of a wave from the equilibrium position.

Analogue signals Signals carried as continuous waves that vary in frequency and amplitude.

Anode The positively-charged electrode.

Artificial satellite A satellite put into orbit from the Earth.

Atmosphere The layer of gases around the Earth.

Atom The smallest part of an element that can exist. Atoms have a nucleus consisting of protons and neutrons around which are shells of electrons.

Atomic number The number of protons present in an atomic nucleus (and the number of electrons present in the neutral atom).

Background radiation The radioactivity that is always present in the environment.

Battery A number of electrical cells joined together.

Beta (radiation/particles) A type of radioactive emission with moderate penetrating power blocked by thin sheets of aluminium.

Beta particles are high-energy electrons.

Big Bang theory A theory that considers that the Universe started from a gigantic explosion.

Black hole An object in space that is so dense and its gravitational field so strong that light and other forms of electromagnetic radiation cannot escape from it.

Braking distance The distance a vehicle travels before stopping after the brakes have been applied.

Capacitor A device designed to store electrical charge.

Cathode The negatively-charged electrode.

Cell (electrical) The single unit from which batteries are made.

Centre of mass The point where all the mass of an object can be thought to be concentrated.

Chain reaction A reaction in which a nucleus is split and neutrons released that can split other nuclei to produce a continuous chain of events.

Charge A feature of atomic particles. Protons and electrons have a charge. Electrons have a negative charge and protons have a positive charge. Opposite charges attract; like charges repel.

Circuit breaker A device that uses the action of an electromagnet to switch off an electrical supply very rapidly.

Comet An object made of ice and rock which orbits the Sun in a different plane to the planets.

Compression (forces) The process of being squashed.

Compression (waves) The region in a longitudinal wave where the vibrating particles of the medium are closer together than normal.

Conduction (electrical) The transfer of electrical energy along a material by free electrons.

Conduction (heat) The transfer of heat energy along a material.

Conductor (electrical) A substance allowing electrical charge to pass through.

Conductor (heat) A substance allowing heat energy to pass through.

Convection The transfer of heat energy in a liquid or a gas (fluid) caused by differences in density. Warmer, less dense fluids rise. Cooler, more dense fluids sink.

Converging lens Any lens that is thicker in the middle than it is at the edges.

Core (Earth) The innermost part of the Earth.

Core (electromagnets) The central part of electromagnet around which current-carrying coils of wire are wound.

Cosmic ray Rays and particles from space reaching Earth.

Coulomb Unit of electric charge.

Crest The point of maximum displacement in a transverse wave.

Critical angle When a ray of light, travelling in a more dense medium, hits the boundary between the more dense medium and the less dense medium and only just emerges by refraction, the angle of incidence of the ray in the more dense medium is called the critical angle.

Crust The surface layer of the Earth.

Cycle (vibration) For a transverse wave one cycle is one trough and one crest. In a longitudinal wave one cycle is one compression and one rarefaction.

Decay (atomic) The break up of unstable nuclei resulting in the production of radioactive emissions.

Decelerate To slow down.

Diffraction The spreading out of a wave as it passes through a narrow gap or moves past an object.

Digital signals Signals carried as a series of 'on' and 'off' pulses.

Diode An electrical device that only conducts electricity in one direction.

Direct current (d.c.) Electric current that does not change direction.

Diverging lens Any lens that is thicker at the edges than it is the middle.

Drag The force from the air that opposes movement.

Dynamo (generator) Device supplying a voltage from the relative motion of a conductor with a magnetic field.

Earthing The linking of a low resistance wire to a metal object to provide a low resistance path to the Earth's surface for an electric current.

Efficiency The ratio of useful energy transferred by device to total energy transferred to device.

Elastic collision A collision that involves no overall change in kinetic energy.

Elastic potential energy The energy stored in an object when work has been done to change its shape.

Electric charge A quantity of electricity.

Electric current The flow of electrons or ions. The rate of transferring electric charge. Units are amperes (amps)

Electrical energy Energy transferred by a charge.

Electrode A negatively or positively charged conductor.

Electrolysis The process of splitting up a chemical compound using an electric current.

Electromagnetic induction The production of a voltage or current across a conductor in relative motion within a magnetic field.

Electromagnetic spectrum The range of frequencies and wavelengths of electromagnetic waves.

Electromagnetic waves Transverse waves that have a common speed in air or a vacuum.

Electrons Negatively-charged sub-atomic particles orbiting in shells around the atomic nucleus.

Electrostatic forces Forces due to stationary electric charge. Like charges repel, unlike charges attract.

Element A substance made up of atoms which contain the same number of protons so contain only one type of atom, and which cannot be broken down into anything simpler by chemical means.

Fetus The name given to an unborn child more than 8 weeks after conception.

Focus The point through which parallel rays of light incident on a converging lens will be refracted.

Fossil fuels The non-renewable energy resources: crude oil, natural gas and coal.

Free electron The electrons in metals that move around inside the metal and do not remain in orbit around a nucleus. The presence of these free electrons allows the metal to conduct electricity and heat.

Frequency The number of cycles (vibrations) per second. Units are Hertz (Hz).

Friction A force which opposes the movement of an object.

Fuse A wire fitted in plugs that is designed to melt if too large a current flows through it.

Fusion (atomic) The joining of small nuclei to form a large nucleus. The process transfers heat energy to the surroundings.

Galaxy A vast number of star systems held together by gravitational forces.

Gamma (radiation) A type of radioactive emission with high penetrating power blocked by concrete/lead. Gamma radiation is part of the electromagnetic spectrum and has a very high frequency.

Geiger–Müller tube (GM tube) A detector of radioactive emissions.

Generator (dynamo) Device supplying a voltage from the relative motion of a conductor within a magnetic field.

Geostationary satellite A satellite which takes 24 hours to orbit the Earth.

Geothermal energy The energy produced in the Earth by natural heating process.

Global warming An international problem caused partly by the increase in the amounts of carbon dioxide and methane in the atmosphere which results in an increase in the average temperature of the Earth.

Gravitational potential energy The energy stored in an object due to the vertical height through which it has been lifted.

Gravity (gravitational force) A force of attraction that acts between all objects.

Greenhouse effect The effect in the atmosphere of heat energy being absorbed by gases such as carbon dioxide and methane.

Half-life The time taken for half a given number of radioactive atoms to decay to different atoms.

Hertz (Hz) The unit of frequency.

Hydroelectric Electrical power generated by the flow of moving water.

Input sensors Devices that detect changes in the environment.

Insulator (electrical) A substance not allowing an electric current to flow and charges to move.

Insulator (heat) A substance not allowing a transfer of heat energy from a hot region to a cold region.

Ion An atom or group of atoms which have lost or gained electrons to become positively or negatively charged.

Ionise To remove or add electrons to atoms or groups of atoms so giving them positive or negative charges.

Isotope Atoms of the same element which contain different numbers of neutrons.

Joule The unit of energy.

Kilowatt 1000 watts.

Kilowatt hour A unit of electrical energy.

Kinetic energy The energy possessed by an object due to its motion.

LDR (light dependent resistor) An electrical component the resistance of which decreases when light shines on it.

Light year The distance a light ray travels in one year.

Lithosphere The outer shell of the Earth made from the crust and the upper part of the mantle.

Logic gate A type of electronic switch used to process information.

Longitudinal wave A wave in which the vibrations of the particles are in the same direction as the energy transferred along the wave.

Magma Molten rock below the Earth's crust.

Magnet An object that attracts magnetic materials such as iron, steel, nickel and cobalt.

Magnetic field The region around a magnet where a magnetic material experiences a magnetic force.

Mantle The layer of the Earth between the crust and the core.

Mass number The total number of protons and neutrons in an atomic nucleus.

Mass The amount of matter an object contains Units are kg.

Milky Way The galaxy containing our solar system.

Moment The size of the turning effect of a force, measured in Nm.

Momentum Defined as mass × velocity. Units are kg m/s.

Moon A natural satellite in orbit around a planet.

Motor effect The motion of a current-carrying conductor in a magnetic field

Neutron A particle with no electrical charge found in the nucleus of most atoms. Its mass is similar to that of a proton.

Newton The unit of force (N).

Non-renewable (finite) energy resources Energy resources that, once used, cannot be replaced.

Normal The line drawn at right angles to a surface.

Nuclear fission The breaking up of a large atomic nucleus to release energy.

Nucleon The protons and neutrons in the nucleus of an atom.

Nucleus (atom) The central part of an atom that contains positively-charged protons and uncharged neutrons.

Ohm The unit of electrical resistance.

Orbit The regular path taken by an object which passes around another object.

Output devices The part of an electronic system controlled by the processor. It transfers electrical enery to other forms of energy.

Parallel circuits Closed electrical circuits that provide several pathways for an electric current.

Pivot A point that objects turn around.

Planet A very large object which orbits the Sun.

Potential difference The voltage between two points in a circuit.

Potential divider A combination of resistors in series, used to split the voltage of a battery into two parts.

Power The rate of transfer of energy. Units are watts.

Primary coil The input coil in a transformer.

Processor The part of an electronic system that decides what action is needed.

Proton A positively-charged particle found in the nucleus of an atom. It has a mass similar to that of a neutron and the number of protons present decides which element is present.

P waves Longitudinal seismic waves which travel through solids and liquids.

Radiation (heat transfer) A process by which heat is transferred.

Radiation (nuclear) The random emission of energy from an atomic nucleus as the result of the breakdown of unstable nuclei.

Radioactive (radiocarbon) dating The use of half-life to date ancient organic objects.

Radioactive decay The emission of particles or rays from an unstable atomic nucleus.

Radioactive emissions The particles and rays produced as the result of the breakdown of unstable nuclei.

Radioactive tracer A radioactive substance, usually with a relatively short half-life, that is passed into the body and used to detect, for example, the presence of cancers, tumours or the direction of blood flow. Tracers can be used in the treatment of cancers and tumours. Tracers can also be used to monitor the flow of liquids and gases in underground pipes.

Radioactivity The random emission of energy from an atomic nucleus as the result of the breakdown of unstable nuclei.

Radioisotope A radioactive isotope.

Radionuclides Materials which produce ionising radiation, such as X rays, gamma radiation, alpha particles and beta particles.

Random Spontaneous not regular.

Rarefaction The region in a longitudinal wave where the vibrating particles of the medium are further apart than normal.

Real image An image that can be shown on a screen.

Red giant A relatively cool giant star.

Red shift The effect on the spectrum of a galaxy moving away from us.

Refraction The change in direction of a wave when it passes from one medium to another due to a change in speed when passing from one medium to another.

Relay An electromagnetic switch.

Renewable energy resources Energy resources that will always be available.

Resistance A measurement describing the difficulty of electric current flow in a conductor. Units are ohms.

Resistor A device for controlling the current in a circuit.

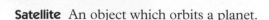

Glossary

Satellite An object which orbits a planet.

Secondary coil The output coil in a transformer.

Seismic waves These are waves created in the Earth by vibrations due to earthquakes.

Series circuits Closed electrical networks giving only one pathway for an electric current.

Solar system A system made up of the Sun, planets, moons, asteroids and comets.

Speed The distance an object travels in a unit of time. Units are m/s.

Star A source of light due to heat caused by nuclear fusion.

Sun A star at the centre of a solar system.

S waves Transverse seismic waves that can only travel through solids.

Tectonic plates The separate slow-moving adjacent sections of the Earth's lithosphere that move because of convection currents within the Earth's mantle caused by the natural radioactive processes within the Earth.

Terminal velocity The constant speed reached by a falling object when the forces acting on it (in the direction of its motion) are balanced.

Thermistor An electrical component in which the resistance decreases when it gets warm.

Thinking distance The distance travelled by a car during the driver's reaction time.

Total internal reflection This takes place at the boundary of two materials when light travelling in the more dense material strikes the boundary at an angle of incidence greater than the critical angle.

Transformer A device that changes the size of an alternating voltage.

Transistor A device that can be used as a high speed electronic switch.

Transverse wave A wave in which the vibrations of the particles are at right angles to the direction of the energy transferred along the wave.

Trough The point of maximum displacement in a transverse wave in the opposite direction to a peak.

Turbine A device that turns a generator.

Ultrasound Sound of too high a frequency to be heard by humans.

Universe Made up of innumerable galaxies.

Velocity The speed of an object in a particular direction. Units are m/s.

Vibration The movement needed to produce a wave.

Virtual image An image that cannot be shown on a screen.

Volt The unit of potential difference 1 volt = 1 joule/coulomb.

Voltage The electrical energy difference of a unit charge moved across two points in a circuit; energy transferred per unit charge.

Voltmeter An instrument used to measure potential difference.

Watt The unit of power.

Wavelength The distance between adjacent crests in a wave equivalent to the distance taken by one complete cycle.

Waves Vibrations that transfer energy but not matter.

Wave speed The distance travelled by a wave in a second. Units are m/s.

Weight The force due to gravity on an object. Units are newtons.

White dwarf A small very dense star.

Work That which is done when a force moves an object a certain distance. Units are joules.

Index

Note: Glossary entries are in bold